Analyzing the Effects of GST on Indian Automotive Sector

by
NARESH KUMAR GOEL

ABSTRACT

Corporate takeover has been a regular phenomenon in the automobile industry and is considered as one of the most important business strategies in today's corporate world, playing a vital role across all business sectors in general and in the automotive sector in particular. Economic pressures, fast technological changes, powerful customer demand, and such other issues drastically influence the corporate strategy of companies in the automobile industry. As a response to that, several automakers engage in strategic alliances, affecting all levels of business processes within the supply chain. This research aims to study the past and present scenario of merger and takeover practices in US, UK, and in India, to examine the impact of the corporate takeover on the growth and financial performance of the selected companies under the automotive industry, to investigate the growth drivers and challenges for the automotive industry, and to identify the future trend of the automotive industry. The findings of the study are based on the primary as well as secondary data.

The current landscape of the automotive industry has evolved due to various corporate takeover and acquisition deals over the last century. The activities of corporate takeovers and acquisitions have evolved over time, becoming wider and more complicated in scope, over the years. They are no longer restricted to mere domestic activity, has now spread across the national boundaries of all the major developed economies. Importantly, Merger and Acquisition (M&A) activities in the automotive industry have enjoyed a far more positive reaction from the stock exchange than other industries, with a merger announcement resulting in an almost instantaneous rise of share prices both locally as well as globally. However, the takeover process also presents enormous challenges to auto suppliers. Clear strategies and efficient post-merger integration are prerequisites for a successful transaction.

The study reveals that there is an impact on the profitability, liquidity and leverage position of the selected automotive companies between the pre and post acquisitions. The study also reveals that there are five important factors which motivate the corporates to go for takeover. These five factors are - value of company, managing growth during the transition period, growth drivers, future plan and geographical expansion, and stakeholders' interest.

It is clear from the results that there is a remarkable rise in corporate acquisitions in India, which is due to business consolidation by large industrial houses, consolidation of businesses by multinationals operating in India, increasing competition against imports and acquisition activities. Corporate acquisitions have begun to gain momentum in India; there are plenty of examples of acquisitions provided in this research, which prove that Indian industries have already entered into the M&A process of value creation through corporate acquisitions. This is especially so for the automotive sector in India. Given its vibrant market, young labor force and low manufacturing costs, India is perfectly poised to take the next big leap in the automotive market. Not only as a rich market, not only with its vibrant young workforce, not only as a target company – but as an international acquirer too. The takeover of Jaguar Land Rover by Tata Motors is a prime example of the same. Business-friendly policies such as make in India and improved ease of doing business are bound to show their positive impacts in the coming few years.

Based on the study it is recommended to conduct in-depth analysis and due-diligence before proceeding for an M&A. Various departments of a company, like, finance, human resource and legal departments, collect information in order to analyse factors that can help in assessing and determining the success of an M&A. These include factors such as rise or decline of revenues, profits, productivity, market share, share prices and so on. Managing cultural differences between companies is an important aspect.

Keywords: Corporate takeover, Globalization, Growing economies, Long-run performance, Clear strategies and effective management strategies, Automotive Industry, Profitability.

TABLE OF CONTENTS

Title	Page No

Abstract iv
Table of Contents vi
List of Tables viii
List of Figures xi

Chapter 1 Introduction 1-14
    1.1 History of Mergers and Acquisitions (M&A) in the Automobile Industry .3
    1.2 Emerging Trends in the Automobile Sector 6
    1.3 Present Global Scenario 8
    1.4 Automotive Industry- M&A Overview 10

Chapter 2: Literature Review 15-34
    2.1 Corporate Takeover - Concept and Practices 15
    2.2 Motives of Mergers and Acquisitions 18
    2.3 Regulatory Framework 21
    2.4 Impact of Takeover and Acquisition 22
    2.5 Challenges in Takeover and Acquisition 24
    2.6 Defence Strategies against the Hostile Takeover 25
    2.7 Hypothesis Formulation 28
    2.8 Research Gap 33

Chapter 3: Regulatory Framework 35-47
    3.1 Legal provisions relating to Takeover 35

Title	Page No

Chapter 4: Research Methodology 49-57
    4.1   Significance of the Study 49
    4.2   Objectives of the Study 50
    4.3   Hypothesis..50
    4.4   Research Design..51
    4.5   Data Collection, Sample Size, and Sampling Techniques.......................51
    4.6   Statistical Tools and Techniques for Analysis...56

Chapter 5: Data Analysis, Results and Discussions 59-124
    5.1   Analysis of Primary Data...59
    5.2   Analysis of Secondary Data...95
    5.3   Hypothesis testing and Results 119
    5.4   Results of Factor Analysis 121

Chapter 6: Conclusion, Recommendations, and Scope for Future
              Research 125-133
    6.1   Conclusion 125
    6.2   Recommendations..129
    6.3   Limitations 132
    6.4   Scope for Future Research 133

LIST OF TABLES

Table No.			Page No
Table 1.1	:	Selected Automotive Industry Transactions	11
Table 1.2	:	Top acquisitions made by Indian companies worldwide	14
Table 5.1	:	Demographic profiles of respondents based on their gender, profession, experiences, and age	59
Table 5.2	:	Respondents views regarding the effect on the value, growth drivers, future trend, and reason with respect to mergers and acquisitions of companies	61
Table 5.3	:	Descriptive Statistics	63
Table 5.4	:	Cronbach's Alpha Reliability Test	64
Table 5.5	:	Wilcuxen t-test based on the views according to the Gender of the respondents	65
Table 5.6	:	Respondents views according to their age regarding the effects of M&A	68
Table 5.7	:	F Test on the profession wise distribution of respondents views	74
Table 5.8	:	F Test on the experience wise distribution of respondents views	80
Table 5.9	:	KMO & Bartlett's test	88
Table 5.10	:	Rotated Component Matrix	89
Table 5.11	:	Total Variance Explained	90
Table 5.12	:	Rotated Component Matrix	91
Table 5.13	:	Results of Factor Analysis	91
Table 5.14	:	Operational and Financial performance of TATA Motors Ltd. during the pre and post takeover period	96
Table 5.15	:	Descriptive test statistics base on profitability ratio- pre and post takeover scenario	97
Table 5.16	:	t-test on Profitability ratios	98

Table No.			Page No
Table 5.17	:	Financial position of TATA Motors Ltd. during pre and post takeover scenario	98
Table 5.18	:	Descriptive test statistics based on liquidity ratios – pre and post takeover of TATA Motors Ltd	99
Table 5.19	:	t-test on Liquidity ratios	100
Table 5.20	:	Descriptive test statistics based on leverage ratios – pre and post takeover of TATA Motors Ltd	100
Table 5.21	:	t-test on Leverage ratios	101
Table 5.22	:	Operational and Financial performance of VOLKSWAGEN during the pre and post takeover period	102
Table 5.23	:	Descriptive test statistics based on profitability ratios – pre and post takeover scenario	103
Table 5.24	:	t-test on Profitability ratios	104
Table 5.25	:	Financial position of VOLKSWAGEN during pre and post takeover scenario	104
Table 5.26	:	Descriptive test statistics based on liquidity ratios – pre and post takeover of scenario	105
Table 5.27	:	t-test on Liquidity ratios	106
Table 5.28	:	Descriptive test statistics based on leverage ratios – pre and post takeover of VOLKSWAGEN	106
Table 5.29	:	t-test on Leverage ratios	107
Table 5.30	:	Operational and Financial performance of MAHINDRA & MAHINDRA Ltd. during the pre and post takeover period	108
Table 5.31	:	Descriptive test statistics based on profitability ratios –pre and post takeover scenario	109
Table 5.32	:	t-test on Profitability ratios	110
Table 5.33	:	Financial position of MAHINDRA& MAHINDRA during pre and post takeover scenario	110
Table 5.34	:	Descriptive test statistics based on liquidity ratios – pre and post takeover of MAHINDRA& MAHINDRA	111

Table No.			Page No
Table 5.35	:	t-test on Liquidity ratios	112
Table 5.36	:	Descriptive test statistics based on leverage ratios – pre and post takeover of MAHINDRA & MAHINDRA	112
Table 5.37	:	t-test on Leverage ratios	113
Table 5.38	:	Operational and Financial performance of DAIMLER CHRYSLER during the pre and post takeover period	114
Table 5.39	:	Descriptive test statistics based on profitability ratios - pre and post takeover scenario	115
Table 5.40	:	t-test on Profitability ratios	116
Table 5.41	:	Financial position of DAIMLER CHRYSLER during pre and post takeover scenario	116
Table 5.42	:	Descriptive test statistics based on liquidity ratios – pre and post takeover of DAIMLER CHRYSLER	117
Table 5.43	:	t-test on Liquidity ratios	117
Table 5.44	:	Descriptive test statistics based on leverage ratios – pre and post takeover of DAIMLER CHRYSLER	118
Table 5.45	:	t-test on Leverage ratios	118
Table 5.46	:	t-test summary of Profitability indicators	119
Table 5.47	:	t-test summary of Liquidity indicators	120
Table 5.48	:	t-test summary of Leverage indicators	120

LIST OF FIGURES/GRAPHS

Figures/Graphs No.	*Page No*

Figure 1.1 : U.S. Automotive Transactions during 2012-2016..............................10

CHAPTER 1
INTRODUCTION

The automotive industry is one of the largest global industries, a key sector of the economy. The industry comprises of the automobile and the auto component sectors and encompasses commercial vehicles, multi-utility vehicles, passenger cars, two-wheelers, three-wheelers, tractors, and related auto components.

The recent changes and developments in the global auto sector have made its impact on the Indian market and manufacturers too. India has made its presence felt in the global auto sector in recent years with the potential for improving profitability. India offers scope for the manufacture of low-cost vehicles with its highly skilled but inexpensive manufacturing base, labor-intensive and developing economy. An important pillar on which the Indian economy stands today is the automotive industry.

In order to achieve rapid growth, Merger and Acquisition (M&A) is a strategic decision so that it generates revenue over the years to come. M&As in India has become a significant part of the Indian economy which is evident as Companies merging or acquiring make headlines daily. The macroeconomic indicators demonstrate the increasing trend of M&A deals.

The automotive industry appeared in Europe in the 17th century and spread across all over the world, from artisan production style passing through "Henry Ford" mass production to the "state-of-the-art" technologies used in present times. Throughout this development process, automakers have been forming alliances with other automakers and suppliers. However, current alliances seem to distinguish in some points to conventional ones such as mergers, acquisitions, and joint ventures.

Corporate acquisition is considered to be an effective strategy in order to acquire market share and reduce competition in case of the automotive industry. Given the changing trends in the global automotive industry, suppliers are under tremendous pressure to carve out M&A deals. The chances of success of M&A deals in the automotive industry are much higher as compared to other industries. The fact is that many of these transactions involve companies of similar business models. Share prices for automotive suppliers jump above average on the mere announcement of M&A activity. Stock markets react just as positively to international and transcontinental mergers and acquisitions as they do to domestic M&A. To ensure the success of M&A deal, it is of crucial importance to select the target company with immense wisdom and have in place some clear-cut post-merger policies. Acquisitions can help companies generate growth, increase their efficiency and competitiveness and boost shareholder value. Many automotive suppliers have realized this in recent years and stepped up their M&A activities.

M&A activities hold out a whole lot of business advantages including improved competitive edge, new market share, lower risks and increased scale of operation. No wonder they are becoming popular across the world.

Better availability of finance, rich corporate and high opportunities for growth may be enlisted as the chief causes behind increased M&A opportunities in India. Besides, it is getting clear that such activities will intensify in the future.

The radical transformation in technology in the automotive industry from the steam engine to gas-powered engine, from fuel to electric engine and from the military tractor and steam-powered tricycle to sedan and hatchback, has paved the way for almost two and a half centuries long journey. In the 19th century, automobile carried the label of rich man's toy because its manufacturing was exorbitant and unaffordable for a bourgeois community. The USA monopolized the automotive industry for many years to come; however, come the end of the second world war around 1945, nations like Japan and Germany began to emerge as potential competitors. By the early 1980s, automobiles of Japanese and German make began to flood the USA markets.

Currently, the automotive industry in the developed nations seems to be experiencing stagnation owing to the poor state in which the automobile market is, in contrast, the automotive industry of developing nations like India and Brazil, where the sales are burgeoning by each passing year.

The contribution of the automotive industry to R&D, innovation, commerce, and Government revenues is more than €430 billion in twenty-six countries. Several manufacturers are among the "Global Top Ten" too.

Globalization is pushing auto majors to come together, diversify their product base, improve technology, hit other markets and, most importantly, economize. Automobile companies are, thus, trying to balance competing priorities while ensuring that they continue to add value. The global recession has reset the automobile industry landscape.

According to experts, the automotive industry is expected to witness multiple M&A deals in the next few years. This will primarily be in the commercial vehicle and two-wheeler sectors. The automobile companies are expected to make large acquisitions to access markets and gain technical knowledgebase.

Rising customer expectations for new technology are putting tremendous pressure on manufacturers to maintain a steady flow of innovative upgradation. Changing demographics and continuing urbanization are altering the mobility needs of the customers. Convergence can help in the development of new and innovative products and services, and thus, with fulfilling the consistently high expectations from the automobile sector.

1.1 History of Mergers and Acquisitions (M&A) in the Automobile Industry

USA Perspective

Starting from the 1890s, the USA M&A deals can be divided into six distinct phases or waves. While horizontal mergers were the commonest ones in the first wave (1895-1904), the second wave was qualified by vertical M&As (1916-1929). Conglomerate mergers characterized the third wave that lasted 1965-1969. The fourth wave (1981-1989) witnessed a number of hostile takeovers. Riding on the wings of growing globalization, the fifth wave showcased a number of cross-border mergers (1990-2000). The wave currently going on (2003-present) has shown evidence of shareholder's activism and leveraged buy-outs.

UK Perspective

Barring two mini merger booms that happened in the 1890s and 1920s, the first merger wave in the UK may be traced to 1968. Here too, the horizontal mergers took precedence, largely sponsored by Britain's Industrial Reorganization Corporation. The second wave was also one characterized largely by horizontal mergers, though a few conglomerate mergers did occur. The third wave of the 1980s was the most substantial one, riding on the wave of the stock market bull run as well as the 'Big Bang Deregulation' of the financial services sector in London. The fourth wave was recorded in the 1990s, with privatizations and deregulations being the striking features driving the deals.

Indian Perspective

Seeds of the Indian automotive industry were seen in the 1880s, but the wave of M&A's came nearly 100 years later, in the 1980s. The acquisitions were usually friendly ones, negotiated through mutual dealings. The Indian automotive industry showed significant advances since de-licensing and opening-up of the sector by the abolition of licensing in 1991, Foreign Direct Investment (FDI) in 1993 and automatic approval allowed foreign investment in priority sectors up to 51% that included the automotive industry, except manufacturer of passenger cars. A recent trend witnessed in the Indian corporate sector is the acquisition of foreign companies by the Indian businesses.

Today, India is in the enviable position of being the largest export base for compact cars to European nations. With the focus on innovation and technology, hybrid and electric cars are not far behind. The manufacturing plants of the world's top-ranking auto-makers in India have been globally acclaimed in terms of their productivity and quality. Giants such as Hyundai, Toyota, and Suzuki established their manufacturing units in India, among other world's leading automotive companies.

The Indian economic reforms since 1991 have given a boost to M&A deals both in the local as well as in global markets. The trends of M&As in India have shown dynamicity in the last few years, having varied across the various sectors of the economy. Today, India stands tall among the other nations of the world as one of the partners of major M&A deals. Recently, many foreign companies have acquired Indian companies as they see a lot of scope in the Indian corporate sector. IT and IT's sectors in India have already proved their potential in the global business.

A number of other sectors in India are also following the trend of resorting to mergers and acquisitions in the recent times, For example, automobile, finance, FMCG, construction materials, steel, and telecom industry. Various key factors like - the entrepreneurs with an aggressive attitude, positive policies by the Government, flexibility in the economy and better liquidity position, have played role in changing trends of mergers and acquisitions in India.

Indian corporate has certainly become confident about successfully spreading their operations overseas. In the last decade, India has emerged as an attractive investment destination, earning $75 billion of foreign investment and becoming one of the 21st largest foreign investors.

The freeing of the industry from a restrictive environment helped it to restructure, absorb newer technologies, align itself to the global developments and realize its potential; on the other hand, this significantly increased this industry's contribution to overall economic growth in the country.

The arrival of most international automobile giants in India has set the stage for exponential growth in the automotive industry's levels of technology, quality, and competitiveness. The simultaneous advent of novel and latest models has accelerated demand for vehicles in the market.

M&A is a significant approach to expansion and development of today's enterprises, which can help rapid access to market channels and expansion of market share, thereby bringing about economies of scale for businesses. Naturally, this restricted the cash flow needed to materialize the transactions and corporate cut down on expansion, both organically and inorganically. This is visible in India as well.

To give a brief idea as to the potential of India - in the automotive industry, the India is one of the most dynamic passenger cars market in the world, the largest manufacturer of motorcycles, the second-largest manufacturer of two-wheelers and the fifth-largest manufacturer of commercial vehicle in the world, and is also becoming a pivotal export center for special utility vehicles(SUVs). The top global automotive manufacturers hope to reap the benefits of India's economic and inexpensive manufacturing practices. In fact, it is possible that India becomes the hub of SUV manufacture and supply, fulfilling the world's need of the vehicles.

Importantly, India is also in the enviable position of being the largest export base for compact cars to European nations. With the focus on innovation and technology, hybrid and electric cars are not far behind. If judged on the basis of quality and productivity, top international brands' manufacturing plants on Indian soil, are among the best.

1.2 Emerging M&A trends in the Automobile Sector

USA Perspective

52 deals were announced in the USA in the year 2016. Strategic purchases made up 96% of the automotive mergers and acquisitions in 2016, this figure was 93% in 2015. Several automotive categories performed outstandingly as per the Standard & Poor's 500 index.

Few of the significant M&A deals executed have been enlisted below:

i. Ningbo Joyson Electronic Corp. agreed to takeover Key Safety Systems Inc. for a whopping $920 million. Funds for the deal were managed by Crestview Partners LP, Canada Pension Plan Investment Board, and Fountain Vest Partners.

ii. Octavius Corporation acquired Grand Design RV for $500 million from Summit Partners LLP and others.

iii. American Axle & Manufacturing Holdings Inc. acquired Metaldyne Performance Group Inc. for a sum of $1.5 billion.

iv. Cruise Automation Inc. was purchased by General Motors for ~$600 million from Felicis Ventures, Signia Venture Partners and Spark Capital Partners among others

v. Trans American Auto Parts Company LLC was taken over by Polaris Industries Inc. for $670 million

In terms of total value the following sectors showed an increase in activity from 2015 to 2016:

i. Consumer products and services
ii. Energy and power
iii. Financial sector
iv. High technology
v. Manufacturing sector
vi. Materials
vii. Real estate

UK Perspective

The United Kingdom automotive industry is opulently rich, with the presence of many global brands. Daimler, Bentley, Lagonda, Jaguar, Aston Martin, Land Rover, Rolls-Royce, Mini, MG, McLaren, Lotus, and Morgan are among the top-notch marquees present here. Among the volume car manufacturers Nissan, Honda, and Toyota are here. London Taxis International, Leyland Trucks, Alexander Dennis, GMM Luton, and Ford are among the commercial vehicle manufacturers.

A few highlights of the UK automotive Industry M&A deals in Q1, 2017 have been outlined below:

i. A total of 127 deals were materialized involving UK-based companies.
ii. 26 successful takeovers were made on foreign soil by UK companies, totaling a worth of GBP1.9 billion.
iii. 40 inward deals worth GBP 5.1 billion were materialized here, with foreign companies purchasing the controlling stake in UK based target companies.
iv. 48 domestic deals were sealed, wherein both the acquirer and the target companies were UK based. Total deal volume was GBP 3.6 billion.

Indian Perspective

Given the challenges of globalization, automotive industry giants are under pressure to diversify their products, improve technology, explore novel markets, reduce expenditure, and ultimately, consolidate. The infrastructure pertaining to testing vehicles for their safety features and emission norms is being refurbished, with financial inputs from the State an invitation to the industry to contribute with technology and expertise. Funds are also being diverted to R&D activities to ensure that vehicles capable of running on alternative fuels, or vehicles better equipped to meet the more stringent emission norms, can be produced. The Government has also responded by lowering the taxation burden. External Commercial Borrowings (ECB) norms have been eased, lending rates within the country have been brought down and other such reforms have ensured that investment has become more conducive. Several non-tariff barriers have been waived as well. In addition, the FDI norms have been relaxed. The Government of India's ambitious plan to connect all major state capitals and metropolitan cities by eight-lane expressways is likely to further boost the automotive industry, especially heavy vehicles.

1.3 Present Global Scenario

Rise in the cross-border activity by India Inc.

Cross-border activities are motivated by a number of factors such as robust cash flows in the country, dynamic global demand, easy availability of finance, demand of novel markets and latest technologies. In order to fulfill these objectives of a business, the processes of M&A's are important.

The Indian scenario and macroeconomics impacting India

The 2016 survey conducted by the Credit Suisse Emerging Consumer Survey showed that India had topped the Emerging Consumer Scorecard, which shows a healthy standard of expectations by the consumer for income. This is indeed distinguishing India in the emerging world.

The positive impact of the Government of India's (GOI) efforts to improve the business scenario in the country is being reflected in India's improved World Bank's doing business ranking in 2016. Especially the ease of 'starting a business' has shown particular improvement.

The recent merger of State Bank of India and its associate banks is likely to quintuple SBI's asset base, leaving behind its competitors. In July 2015, a PIB (Press Information Bureau) released by the Ministry of Commerce and Industry, GOI, confirmed a 48%growth in FDI equity inflows marking the success of the "Make in India" campaign. Clearly, investors from around the world are showing their faith in an upcoming Indian economy.

The drive of demonetization by the Government affected every individual in the country and the economy as a whole, to a large extent. It was a committed effort to integrate both, unorganized and organized, sectors. With the surge in fund deposits in banks is an increased demand of loans in the high-growth sectors. This will bring down the interest rates, leading to various benefits such as lower cost of production, higher profits and diversified growth. These favorable reforms in the economy could have ripple effect, leading to better availability of resources for further development of the country future plans, including smart cities.

India now ranks 39 among 138 countries in the World Economic Forum's Global Competitiveness Index which ranks countries on variables such as institution's education, market size, infrastructure and macroeconomic environment.

The Indian automobile industry is among the world's largest and one of the fastest globally growing industry. India reported yearly manufacturing of more than 3.7 million units of passenger cars as well as commercial vehicles, propelling the country to the position of being the seventh largest manufacturer in the world. India stands as the fourth largest exporter of passenger cars in Asia - behind Japan, South Korea, and Thailand.

M&A activities Worldwide

Latest statistics and trends: Given the current scenario of uncertainty and market dynamics, mere organic growth in the automotive industry is not a possibility anymore. The sources reveal an 18% annual decrease in global mergers and acquisitions in the year 2016, from USD 4.66 trillion in 2015 to USD 3.84 trillion in 2016. The outbound volume for China witnessed a record high at USD 225.4 billion. USA in bound M&A value was also high at USD 486.3 billion. In October 2016, the global mergers and acquisition on record were 600.8 billion USD.

In the first three quarters of 2016, M&A activity rose in India, with 712 deals in total and an increase of 135 deals year-on-year basis, according to the EMIS (a Euro Money Institutional Investor Company) report on Asia markets.

Macroeconomic trends: Global economy thrives on emerging as well as mature markets. Currently, the trend is of emerging markets investing in developed ones, which can have stupendous effects on future deals. Financial institutions have more money at disposal, thanks to the monetary easing policies in developed countries, such funds are being directed to merger and acquisition deals.

Capital market of an economy plays an influential role in M&A activities. The decision by the Federal Bank of USA will probably affect worldwide capital markets, leading to volatility before things stabilize. The uncertainty is deepened with occurrence of incidents such as Brexit, the consequences of which cannot be measured yet.

1.4 Automotive Industry - M&A Overview

USA Perspective

U.S. automotive industry transaction activity has remained consistent in recent years. The following graphs show the summary of M&A transactions during the last five years.

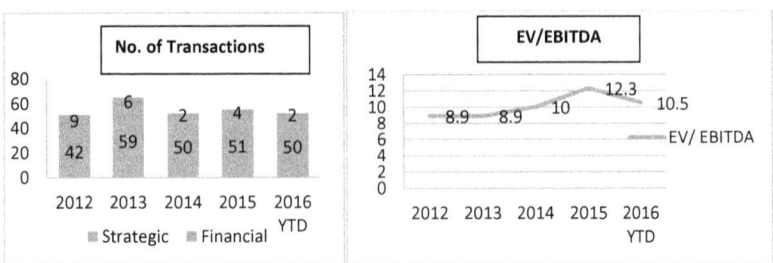

Source: Adapted from Capital IQ, 2017

EV-Enterprise value.

EBITDA- Earnings before interest, tax, depreciation, and amortization.

Figure no. 1.1: U.S. Automotive Transactions during 2012-2016

The following table shows some selected automotive industry transactions in the US during the year 2016.

Table no. 1.1: Selected Automotive Industry Transactions

Date	Target / Issuer	Transaction Size ($mm)	Buyers / Investors
02/12/2016	Bomnin Chevrolet West Kendall	$21	Bomnin Chevrolet
03/11/2016	Metaldyne Performance Group Inc.	$3,418	American Axle & Manufacturing Holdings, Inc.
27/10/2016	Atwood Mobile Products LLC	$13	Lippert Components, Inc.
16/10/2016	AC Propulsion, Inc.	$100	Chongqing Sokon Industry Group Co., Ltd
13/10/2016	NOCO Energy Corp., Lubricants Business	$37	Brenntag AG
12/1202016	Transamerican Auto Parts Company, LLC	$665	Polaris Industries Inc.
03/10/2016	Grand Design RV	$503	Octavius Corporation
03/10/2016	Remy International, Inc	$80	Torque Capital Group LLC
28/09/2016	Irvin Automotive Products, Inc.	$175	Piston Automotive, LLC
21/09/2016	Corby, LLC	$1	US Lighting Group, Inc.
16/09/2016	Interstate Ford, Inc.	$6	Cincy Autos, Inc.
05/07/2016	Vacuplast, LLC	$11	Patrick Industries, Inc.
01/07/2016	Jayco, Inc	$582	Thor Industries, Inc
28/06/2016	General Products Corporation	$12	AAA-GPC Holdings, LLC
23/05/2016	General Cable Corporation	$71	Standard Motor Products, Inc.
08/04/2016	Quantum fuel Systems Technologies Worldwide, Inc.	$25	Douglas Acquisitions, LLC
11/03/2016	Cruise Automation, Inc.	$578	General Motors Company
29/02/2016	Federal-Mogul Holdings Corporation	$282	IEH FM Holdings LLC
29/02/2016	Pittsburgh Glass Works LLC	$635	LKQ Corporation
26/02/2016	DYK Automotive, LLC	$137	The Sterling Group, L.P.
24/02/2016	Prestone Products Corporation	$230	KIK Custom Products, Inc.
17/02/2016	Metropolitan Automotive Warehouse, Inc.	$23	Parts Authority Metro LLC
16/02/2016	AFX Industries L.L.C.	$77	Exco Technologies Limited
02/02/2016	MICO, Incorporated	$75	
02/02/2016	Key Safety Systems, Inc.	$920	Ningbo Joyson Electronic Corp.

Source: Adapted from Capital IQ, 2017

UK Perspective

As per the office for National statistics 2017, between 1997-2001, 128 completed acquisitions were reported, the mean value of a single deal being £160 million. However, in the coming decade, this number decreased to 67 only; not only that, the average cost per deal went down too, touching £119 million. This further declined to 30, though the mean value of a deal went up to £141 million. To further summarize, if one looks at the 20-year period between 1997-2016, the mean number of outward acquisitions was 76 and the mean volume of the deal was £130 million, In the first quarter of 2017 (Jan-Mar), 26 outward acquisitions had been concluded at the rate of £72 million. The corresponding figures for inward acquisitions between 1997- 2001 were 54 and £182 million per transaction. The next ten years showed remarkable stability in terms of M&A deals. However, this trend was not maintained in 2012-16, with the number of deals falling to 41. Nevertheless, the average deal cost went up to £351 million. This higher average can be attributed to a few very large deals to the tune of £1000 million, which were executed in 2016. The corresponding 20-year average for the period 1997-2016 was 49, with the average per transaction value being £232 million. In Quarter 1 of 2017 (Jan-Mar), 40 deals of inward acquisition were sealed, with an average cost of £129 million per transaction.

Experian, a global information services company stated, in its United Kingdom and Republic of Ireland M&A Review for Quarter 1 2017, that: "After a busy 2016, characterized by increasing volumes in the SME space and robust growth across a range of sectors, the first quarter of 2017 presents more of a mixed picture in terms of UK M&A. While the subdued volume figures suggest that dealmakers have been content to largely hold their fire in Q1, some British PLC's have believed the wait and see approach and pushed ahead with ambitious plans for growth by acquisition, and a number of big-ticket acquisitions mean that overall value figures still compare favorably to the first quarter of last year."

Bureau Van Dijk, a leading information provider of private company, corporate ownership, and deals Information, stated in its Quarterly M&A Review on Report, Global Quarter 1, 2017, "that the volume and value of global mergers and acquisitions (M&A) declined in the first quarter of 2017 in comparison both Q4 2016 and Q1 2016."

Indian Perspective

The automotive sector in India, comprising of the automobile and auto component sub-sectors, is one of the key segments of the economy, having extensive forward and backward linkages with other key segments of the economy.

Government initiatives and policies: Many foreign companies are setting up their facilities in India on account of various Government initiatives like 'Make in India' and 'Digital India.' The aim of the 'Make in India' initiative is to boost the Indian manufacturing sector, which would further boost demand, and hence spur development, in addition to benefiting investors. In addition, the 'Digital India' initiative of the Indian Government emphasizes upon three core components – (i) creation of digital infrastructure, (ii) delivering services digitally, and (iii) increase digital literacy. Currently, the manufacturing sector in India contributes over 15% of the GDP. The aim of the Indian Government is to raise this contribution to at least 25% of the GDP.

Also, various Indian Government initiatives are underway to ensure state-of-the-art infrastructure. The Road Transport and Highway Ministry have set an ambitious target of adding 25,000 KMs in FY17, out of which 15,000 KMs would be awarded by NHAI and the remaining 10,000 by NHIDL. In FY 2015-16, the ministry awarded 10,000 KMs of highway worth INR one lac crore. Better inter-state connectivity is bound to bring about an upswing in transportation. This will subsequently create a demand for personal and commercial vehicles. The growth of the automobile sector has been possible through the Government of India's Automotive Mission Plan (AMP) 2006-16. It is being predicted that this sector's share to the GDP will reach US$ 145 billion in FY17 due to the increased focus on exports of two and three-wheelers, small cars, multi-utility vehicles (MUVs) and auto parts by many Indian companies.

Investment: India is an emerging global hub for sourcing automotive vehicles. Geographically, it is closer to key automotive markets like the ASEAN, Japan, Korea, and Europe. The Government of India encourages foreign direct investment (FDI) in the automobile sector and allows 100% FDI under the automatic route.

Positive economic indicators: Macro-economic indicators such as GDP, Inflation, REPO rates, and consumer sentiments have been quite positive in recent times. It is expected that interest rates will further come down from the current level. These favorable key indicators are expected to drive consumer spending, which in addition to the availability of credit, should positively impact the demand for automobile and auto parts.

Table 1.2: Top acquisitions made by the Indian companies Worldwide

Acquired Company	Acquirer	Deal Amount	Date of Deal
Corus Group (U.K.)	Tata Steel U.S.	$12.11 billion	January 31, 2007
Zain Africa	BhartiAirtel	$10.7 billion	February 2010
Novelis (U.S.)	Hindalco Industries	$6 billion	Feb 11, 2007
Jaguar Cars and Land Rover (U.K.)	Tata Motors	$2.3 billion	2008
Honiton Energy Holdings (China)	Tanti group	$2 billion	April 2010
Abbot Point Coal Terminal (Australia)	Adani Enterprises	$2-billion	May 2011
Algoma Steel (Canada)	Essar Steel Global	$1.85 billion	April 2007
Marcellus Shale (U.S.)	Reliance Industries	$1.7 billion	April 2010
Minnesota Steel (U.S.)	Essar Steel Holdings	$1.65 billion	April 2007
Wockhardt	Negma Laboratories	$265 million	2007
Imperial Energy Plc.	Oil and Natural Gas Corp.	$1.9 billion	2009
Bennett Coleman & Co	Virgin Radio	$53.2 million	2008
Oil & Gas Assets (Kashagan oilfield)	ONGC	$5 billion	November 2012
Port Terminals (Abbot Point X 50Coal Terminal) Australia	Adani Enterprises	$1.97 billion	May 2011
Orient-Express Hotels (Bermuda)	Indian Hotels Co	$1.67 billion	October 2012
Empresa Mixta. (Venezuela)	ONGC Videsh, Oil India, RepsolYPF, Petroliam NasionalBhd-Petronas and Indian Oil Corp	$1.54 billion	February 2010
Oil & Gas Assets (Campos Basin Oil Fields)(Brazil)	Oil & Natural Gas Corp	$1.4 billion	January 2006
Oil & Gas Assets (Eagle Ford Shale gas field) (United States)	Reliance Industries	$1.35 billion	June 2010

Source: Adapted from Annual Report of the Ministry of External Affairs, 2013

CHAPTER 2
LITERATURE REVIEW

This chapter purposes to provide an overview of the existing literature that chains the objective of this research work. It also helps in understanding the theoretical root and to present the numerous viewpoints offered by different studies on the impact of the corporate takeovers on the automobile industry. Literature review discussed is from the published material in a specific area and sometimes information in a specific area within a certain time period. The assessment of the past studies in the research means to note the observations, exploration and numerous more things completed in the past. It offers details regarding tools used, procedures accepted, findings and observations made. It also hints the way for data collection and introduction from the numerous sources. This kind of study is also valuable to define the scope of our research. It saves time, vigor and supports indirectly towards a specific goal. The chapter is fragmented into numerous sections. This chapter also discusses the previous work pertaining to corporate takeovers and helps to develop a theoretical base of the study.

2.1 Corporate Takeover - Concept and Practices

Chandratre (2010) defined takeover as succeed to the management or ownership or take control of; the assuming of control, ownership or management of a corporation. In other words, it is a process of acquisition of a going business by another through outright purchase; the buying of one company or most of the shares in it, by a person or another company, to buy a company or gain control of it, by buying shares in it from the shareholders. Thus, takeover refers to the acquiring control of an existing company, through buying or exchanging shares.

Carleton and Lineberry (2004) stated that an M&A activity will succeed only if apart from financial issues, cultural and social issues have been given adequate attention. Employees will appreciate if their personal issues are taken care of.

Zarin and Yang (2011) reported that M&A activities can be friendly or hostile. Hostile bids are takeover operations wherein the acquiring company gains supremacy over the target company without the consent of the latter's board of directors or top management. The takeover is affected through the persuasion of the shareholders of the target company to part with their stock. Every publicly listed company can be the potential victim of a hostile takeover. To combat this there are a host of defense mechanisms, some proactive (Poison Pill and Staggered Board) while others are reactive (Crown Jewel and White Knight).

Ross, Westerfield and Jaffe (2005) reported that horizontal, vertical and conglomerated acquisitions are three types of takeover transactions by way of which a bidder company acquires control over a public-listed company. There are various ways in which this deal can be executed. There can be a group of external investors or board members driving the deal, purchasing the target company lock, stock, and barrel. A second way of doing so is by an individual shareholder convincing the others to forfeit their voting rights to him, thereby this individual stakes his claim upon the target company by way of voting rights. The second method is, of course, fraught with difficulties as it will take a very good reason for shareholders to be convinced to give up their voting rights. The third option is through the path of M&A. They also enlisted various reasons behind a takeover bid. The clearest reason is to improve market shares, enhance productivity, or increase the stability of positive results. The type of acquisition is decided by the purpose behind the takeover bid. For instance, if the aim is to gain market shares in hitherto untouched areas, the takeover bid is a horizontal acquisition. If the aim is to get better productivity, it is a vertical acquisition. If, on the other hand, the aim is to increase the stability of positive results, the type of acquisition involved is conglomerate.

Weston, Mitchel and Mulherin (2004) discussed that in the case of a vertical acquisition, one company takes over another company operating in the same industry as the acquirer. The major aim behind this is the acquiring company's desire to have complete ownership over the entire production chain, thereby making its market position secure. A good example of vertical acquisition is the takeover of a travel agency by an airline company. By doing this, the airline company gains better control over its potential customers and cuts off competition. They also described that in the case of a horizontal takeover, two companies, that are competitors in the same industry, are involved. As in other cases, here, too, the motivating factors behind the merger are manifold. However, the most fundamental aim is to go for advantages in economies of scale by ensuring improved leadership of the company and by taking advantage of the hitherto unused but available production capacity. Horizontal acquisitions have yet another impact–albeit a negative one on the competition. This is so because it increases monopolistic tendencies in the market, as more and more companies are taken over. This type of take-over bid is fairly common since it is usually by one company to attempt to grow big by acquiring another operating in the same industry. If a company is looking to cover the base in a geographically distinct and unknown market, it makes better business sense for it to attempt to buy out a company with popular brand value and one that is already established in this geographical region. In such events, it is not uncommon for the newly merged entity to opt for a new brand name, one that is preferably reminiscent of both the parent entities. This nomenclature sort of underscores the merger in the public eye and is also a tool to assure the existing customers of the acquired company of continued quality of products or services. A conglomerated acquisition is said to have taken place when one company acquires a target company that hails from a totally different industry. Such acquisitions are aimed at risk-dilution, and often those companies opt for it which is operating in a high-risk industry. The aim is to have fingers in different pies, and ensure over-all profits even if one's own industry is having a rainy day. This approach indeed increases profit sustainability.

Grabianowski (2005) examined the reasons for hostile takeovers. There are several reasons why a company might want or need a hostile takeover. They may think the target company can generate more profit in the future than the selling price. In a strategic acquisition, the buyer acquires the target company because it wants access to its distribution channels, customer base, brand name, or technology. In some cases, purchasers use a hostile takeover because they can do it quickly and they can make the acquisition with better terms than if they had to negotiate a deal with the target's shareholders and board of directors.

2.2 Motives of Mergers and Acquisitions

Krishnamurti and Vishwanath (2010) stated the need for takeovers. The market for takeover is a market in which alternative owners (bidders) compete for the rights to manage (under-performing) companies. The shareholders of the target company can exercise the choice of selling their shares to the highest bidder. Further, large companies are typically owned by a large number of small investors who do not have the incentive to monitor managers' performances because of which managers may get away uncontested. In the absence of a control mechanism, managers waste resources. The job of the market for corporate control is to discipline erring managers. From managerial to financial, there may be several motivations for opting for an M&A deal. Differential efficiency theory underscores the differential efficiencies of different management of different companies.

Manne (1965) highlighted that there was a positive correlation between the market price of the shares of a company and the efficiency of its management team. Clearly, the market price of a company will increase if its management changes hands from a less to more efficient ownership. It becomes an attractive takeover if the acquiring company believes that they can be better and efficient managers of the company. Companies with similar businesses operating in the same sector are more likely to be the acquirers since will have a better understanding in order to have good capability to spot low performance and will have ability and expertise to turnaround the company.

The theory of 'Inefficient Management' is connected to the theory of 'Differential Efficiency'. An acquisition may be perceived as the shareholders of the target company attempting to bring about a positive change in the managerial efficiency of their company. Several times managers of the target company have continued with their traditional practices despite the out datedness or proven inefficiency of the same. The shareholders tend to punish such recalcitrant management through an effective takeover bid.

Jensen and Ruback (1983) stated that a premium should be paid to the holder of the share over and above the current market price, in order to convince shareholders to sell their shares.

One may aim at attaining "Operating Synergy" through horizontal, vertical as well as conglomerate mergers. The assumption behind this theory is that economies of scale are present in the industry and before a merger, the level of activity at which the individual companies are operating, is lower than the potentials for economies of scale. Synergies are achieved in four ways: cost, revenue, market power, and the intangibles. Cost synergies are of two categories namely, fixed and variable cost synergies. Fixed cost synergies include sharing services which are common to all, such as accounting and finance, the office, executive and higher management, legal, sales promotion and advertisement, lower overhead costs to a large extent. Reducing variable cost involves better productivity and stronger buying power. Synergies of revenue are associated with cross-selling products or services through complementary sales organizations or distribution channels that sell different geographic regions, customer groups or technologies. Among intangibles are such things as brand name extensions and knowhow sharing. Such synergy is made possible by transferring the intangible know-how capabilities from one company to another.

According to Thanos and Papadakis (2012) the main benefit of accounting-based measures is that they measure actual performance based on annual financial statements and do not base their performance measure on investors' expectations.

Two hypotheses given by Chatterjee and Meeks (1996) are – capital markets are not completely effective, and the informational efficiency of capital markets is overvalued and concludes that it is expedient to evaluate post-acquisition performance by using accounting-based measures. An additional advantage of these measures compared to stock-based measures is that they can measure the success of Companies that are not listed on the stock exchange. At times, the board of directors or other significant decision-makers hold fewer shares, in which case, the takeovers are aimed at benefiting these key individuals rather than the other shareholders. Benefits include financial as well as non-financial incentives such as better salaries or improved status in the power structure.

As per Jensen and Meckling (1976) if boards have major shareholdings, then the board of directors will be bearing a significant proportion of these costs and will not waste combined wealth through takeover bids designed for selfish purposes. Thus, the purpose of board share-ownership is to ensure that the incentives of shareholders and directors are not at cross-purposes.

Cosh, Hughes, Lee and Singh (1989) opined that a takeover bid could create more value when the target companies had mechanisms in place to counter-balance managerial power (such as off-board institutional shareholding), as compared to the companies that did not.

Fama and Jensen (1983) opined that if offsetting shareholdings were not an option it is the top management who can avoid the takeover bid or dismiss the same through shareholder voting. After all, the top management is expected to own a large share of the company's equity and possess sufficient influence.

It was inferred by Chatterjee and Meeks (1996) that experimental studies on the financial impact of acquisitions are based primarily on two statistical evidence sources – stock market data and accounting rates of return. Since both the measures are expected to reflect post-takeover cash flows it might be expected that, when assessing the same sample of takeovers, they would lead to broadly consistent results.

Bild, Guest and Runsten (2005) investigated whether an acquisition leads to financial benefits for the acquiring company. A cost-benefit analysis taking into consideration the immediate pay-out value is critical. Several other studies have targeted this aspect, majorly adopting two methodologies – 'Profitability studies' and 'Share return event studies'. In 'Profitability studies' there is a straight forward comparison between post-acquisition and pre-acquisition performance of the acquiring company. In the case of 'Share return event studies' there is a study of the impact of the takeover on the share price of the acquired and acquiring company. The profitability studies are conducted by comparing the performance of the acquirer and acquired companies before the takeover bid (combined or weighted average) and after the takeover bid (when the acquired company also becomes a part of the acquirer) then this difference is compared with a benchmark value, which in turn is based on control companies as per industry and volume.

Generally, the studies on takeover deals with return on the equity to the acquirer and acquiree. Since these studies measure returns over very short time periods, compared to profitability studies, they have the advantage of being less subject to problems of noise and benchmark error. The results show huge gains for target shareholders, zero to negative returns for acquiring shareholders, and significant gains overall.

Shleifer and Vishny (2003) concluded that acquisitions (especially involving equity) may declare to the market that the acquirers' company is overvalued, and part of the announcement (and long run) return may reflect a negative reaction to perceived overvaluation rather than fundamental value destruction.

2.3 Regulatory Framework

According to Das and Mishra (2013) M&A in India have been guided by the age-old takeover rules. The Securities and Exchange Board of India (SEBI) realized the need to revamp these rules to keep them in tandem with the continuously changing global scenario. On September 2011, the SEBI amended the new set of takeover rules i.e.; the SEBI (Substantial Acquisition of Shares and Takeovers) Regulations, 2011. The main purpose is to prevent hostile takeovers and at the same time, provide some more opportunities of exit to innocent shareholders who do not wish to be associated with a particular acquirer. In the wake of these rules, promoters as well as public shareholders of a public-listed company shall be eligible to receive the equal price for their stocks. In another shareholder-friendly move, SEBI has scrapped the non-compete fee or control premium, which were being paid to only the promoters earlier and could have been as much as 25% of the public offer price. The SEBI has successfully done one part of the reform process by preparing the new takeover code; the other part requires its successful implementation.

Machiraju (2007) stated that corporate ownership and restructuring assume many forms. The traditional approach through horizontal and vertical mergers to realize synergy has been extended to cover takeovers, management control and change in ownership structure of the company. Business combinations, corporate restructuring, mergers, acquisitions, takeovers are rather significant milestones in the growth trajectory of a business organization. Generally, merger represents a process of allocation and reallocation of resources by companies in response to changes in economic conditions and technological innovations. Takeovers will be more common in future and it is therefore desirable to have in place a code which will be equitable to all the parties, flexible enough to deal with complexities and simultaneously protect the interests of the investor.

As per a report by Price water house Cooper (2017), a takeover of shares is allowed only with the approval of the audit committee and the board of directors, taken in advance. This is as per the Company Law. Sale of shares between companies may also require shareholders' approval in advance. As per procedure, a scheme needs to be approved first by the audit committee, then by the board of directors, followed by the stock exchanges and the shareholders. The deal is to be approved with the required majority, a majority in number plus 75% in value of shareholders or creditors voting personally, or through postal ballot or proxy. Final approval is sought from the National Company Law Tribunal (NCLT), the Court, which does so after seeking opinions from a host of other authorities which includes SEBI, Income Tax department, Regional Director, Registrar of Companies, Official liquidator, RBI, Stock Exchanges and Competition Commission of India. In addition, it also considers the objections, if any, filed by any affected shareholder.

2.4 Impact of Takeover and Acquisition

Several research studies have attempted to answer the million-dollar query – do mergers lead to tangible financial benefits for the bidding Companies. As per Lubatkin (1983) the answer seems to be "Yes". Most of these studies have targeted M&A deals in the USA, Europe, and Australia, with Asia represented by India and China.

Harari (1997) studied how, specifically, cost efficiency is affected by bank mergers in the Taiwanese banking industry. They gathered that there is a positive relationship between bank merger activity and cost efficiency. As per authors such as Hopkins (1999), Peng and Wang (2004), Epstein (2005), and Duncan and Mtar (2006) takeover bids can lead to enhanced cost efficiency. Hence, the next logical query is – how come some M&A deals are more profitable than others. That is to say, what decides the hierarchy of or the degree of success of an acquisition deal. A number of times attempts are made to find answers through case studies. In many ways, the process of M&A's is paradoxical. Stahl and Mendenhall (2005) concluded that an M&A deal would fail if it could not create value for the shareholders. For instance, the USA banking sector M&A's did not result in improved financial efficiency. They also gave the example of General Electric as having grown exponentially due to well-executed acquisition deals. Getting access to new products, new markets, new networks, and new brands seems to be the underlying factor motivating M&A deals. Similarly, Vanitha and Selvam (2007) concluded that the efficient management system of the acquiring company was the reason behind the continued success of the newly merged entity.

Azhagaiah and Sathish (2011) demonstrated that in India corporates from the manufacturing sector gained financially from the merger deals. These companies enjoyed better performance, higher profits, enhanced liquidity position, and better risk management.

Pawaskar (2001) in their study paper titled the "Effect of Mergers on Corporate performance in India" studied the comparison between the operating performances of companies entering into M&A deals, between 1992-1995, before and after they had entered into the agreement. In this study, they outlined the profit profiles. A remarkable finding from the regression analysis was that there were no major profits emerging from the merger deals. However, the merged entity did perform better than the industry performance.

Ramaswamy and Waegelein (2003) in an analysis "Company Financial performance following Mergers" concluded that post-merger performance is significantly positive; there is a strong correlation between differences in the size-structure of the participating companies and the merger success. Usually, if the bidder company is smaller in size than the target company, the post-merger period can be ridden with challenges. Conglomerate mergers were found to meet better success than same industry mergers. Also, companies opting for long-term compensation plans showed better financial performance post-merger. There was also an element of time. Deals executed before 1983 were more profitable as compared to those between 1983-1990 based on performance in the long run but pointed out reasons behind such performance.

Jackson (2015) stated that automobile sales are on the rise globally, and the automotive industry remains one of the most popular sectors for merger deals. Analysts predict that the surge in deal flow that occurred during 2014 will continue or even accelerate throughout 2015. In addition to increased automobile sales, growth and diversification objectives, and macroeconomic factors, deal activity is being spurred by stringent efficiency standards that are pushing manufacturers to reduce vehicle weight. In response, many automotive suppliers are pursuing transactions to acquire new technologies and advanced materials.

2.5 Challenges in Takeover and Acquisition

Krishnaveni and Vidya (2015) outlined what challenges an acquiring company may face during a takeover bid. The Indian automotive industry is particularly plagued by poor roads, inefficient maintenance, and insufficiency of the national highway system, which is barely 2% (or less than that) of the total road length. Even existing roads are old and not wide enough, badly crowded with two-wheelers, pedestrians, and even cattle. Poorly enforced traffic laws and abysmal road infrastructure are two major deterrents to the automotive industry in India. Despite heavy investment and involvement of private players, better road infrastructure remains a challenge in India given the heavy costs and land acquisition challenges. In addition, localization will be an important consideration for foreign investors in India given the multiple challenges associated with import.

Bebchuk, Coates and Subramanian (2002) explained that the negative attitude of the board of directors of the target company may be caused by dissatisfaction over the volume of the bid, though this may be only one of several reasons. Their personal fears of imminent job loss in the near future may be a personal reason, whereas thinking from the company's perspective, the board of directors may feel that the acquisition will not be good for the target company's growth, strategy or revenue.

The growth potential of the automotive industry in India is immense, given the current penetration rate of merely 7 (seven) cars per 1000 (thousand) consumers. A large and sustainable boom in vehicle purchase seems undeniable, given the enhanced purchasing power of nearly 300 million Indians to the range of $1000 per capita income. The reported figure is 1 million to 3 million passenger vehicle sale from 2003-04 to 2015 and volume increase from $9.8 billion to $15 billion from 2005-06 to 2015. Clearly, the degree of competitiveness in this industry is governed by the individual company's capability to improve and innovate. It is certain that the industry will benefit if competition is strong in the local market, strong customer base, and local suppliers. The factors that determine competitiveness are economies of scale, labor charges and duties and interest rates. Even stronger factors are productivity and capacity utilization.

The Indian markets showcase that despite high tariffs being deterrents to trade, a high volume; the positively growing market will still cause investments to flow inwards. However, the Indian manufacturing scenario is not as attractive as other Asian options like Indonesia, Thailand or China. Fuel prices have been known to impact driving preferences of consumers. Fuel-efficiency is one significant factor controlling what type of car the consumer will prefer. For instance, they may prefer cars running on CNG, LPG or diesel or go for smaller, more mileage-efficient vehicles. The same holds true for the commercial vehicles.

2.6 Defence Strategies against the Hostile Takeover

Pearce and Robinson (2004) described that in the face of a hostile bid, the target company can choose from a set of defense strategies, also called anti-takeover measures or shark-repellent tactics. The main purpose of defense strategies is to make the deal unpalatable for the bidding company, especially financially. Also stated that a proactive measure is employed to ensure that the target company becomes less attractive even before the actual hostile bid is made; the next one is brought into practice with respect to the hostile bid.

Bebchuk, Coates and Subramanian (2002) explained that "Staggered Board" is a solution for an acquiring company planning to take control over a target company. It is a means by which the acquiring companies aims at gaining representation on the board of the target. Clearly, through this approach, the acquirer company can better influence the other directors and shareholders. To formulate a staggered board, the shareholders need to approve it during a shareholders meeting. This is also the way to get a seat on the board of directors and can be managed by purchasing voting power through the sufficient number of shares, enabling one shareholder to become a member of the board of directors. The staggered board approach ensures that the task of taking control over the board becomes a lengthy and economically infeasible process.

Ruback (1988) explained that decision-making with respect to M&A deals needs more than 50% of the votes for the decision to be approved. However, in the event that a company includes a "Super-Majority" amendment in its corporate manual then the requisite percentage for approval rises to 67-90%. This amendment of super-majority can be brought in by the shareholders, though it can be activated by the board of directors. Hence, although the acquirer can still attempt to a takeover by submitting a merger proposal to the shareholders, he will require a higher degree of acceptance.

According to Lambert and Larcker (1987), "Golden Parachute" is an effective anti-takeover strategy as it makes the deal more expensive for the bidder by making it compulsory for the top executives to be paid a handsome amount in the event of a takeover in exchange of their job. As soon as threshold values (usually 26.6%) of target company shares have been acquired, the golden parachute strategy gets activated. The strategy is usually not employed by itself but in combination with others. A survey conducted by Lambert and Larcker (1987) revealed that the golden parachute anti-takeover defense strategy could enhance stakeholder wealth by an average of 3%.

Walking and Long (1984) supported this logic, deducing that the top management will oppose an acquisition bid depending on the effect of the takeover on their personal financial status. Hence, the importance of the golden parachute strategy is to ensure that the top management and the shareholders are not working at cross-purposes and their fiduciary interests are aligned.

As per Weston, Mitchell and Mulherin (2004) the board of directors should attempt to show the shareholders how technically incorrect the hostile bid is. They should clarify how the bid value is too low (which it usually is), how it is not a true representative of the worth of their company, and how dim the future of the merger can be. Thus, the board of directors can raise doubts in the minds of the shareholders with respect to the hostile bid, encouraging them to wait for a better offer. Clearly, this is the most cost-effective strategy. However, it can backfire as the shareholders may feel that the board of directors are motivated by personal gains and may not have the best interest of the organization in their hearts. Alternately, the board of directors can pitch another potential bidder so as to encourage higher offers.

Weston (2001) mentioned that several times a hostile bid is aimed at the target company's asset or operations. The "Crown Jewel" defense strategy ensures that the target company can outright sell all or some of the company's best assets to make itself less attractive as a prey. Alternatively, the target company may sell its crown jewels to a friendly bidder – termed as the white knight. Later, when the hostile acquirer has withdrawn its offer, the target company can buy back the crown jewels from the white knight at a pre-fixed cost.

Weston (2001) outlines the major characteristic of a "White Knight" and "White Squire" mechanism as the requirements of a third entity. Firstly, the target company finds itself a sympathetic company that steps in as the white knight and purchases a majority of shares in the target company. Through this, hostile takeovers can be avoided. The white knight is chosen because of a deeper faith in intentions, or a better history of the relationship, with the chance that the target company retains its independence. Even if this independence is lost, the target company has still managed to escape the clutches of the hostile bidder. At the hands of the white knight, the target company can hope of a better, more secure future. The other purpose achieved by the white knight approach is to offer competition to the acquirer company, tempting it to up the bid. However, whatever be the purpose, the challenge is to find a white knight in the first place, as it is not too easily available an option.

Liberalization policy was adopted by the Indian Government from 1991 onwards. In 1994, SEBI came out with the rules and regulations namely "Substantial Acquisition and Takeovers Regulations". These regulations contain the process to be followed by acquirers while acquiring a stake in the Indian companies.

2.7 Hypothesis Formulation

A research titled "A Discriminant Analysis Function for Conglomerate Targets" by Simkowitz and Monroe (1971) used "Multiple Discriminant Analysis" (MDA) to analyze conglomerate target companies that merged in the year 1968. Data was taken for 25 non-merged companies and 23 merged companies from the COMPUSTAT tapes in order to build the discriminant function. A holdout group of 23 companies was used to test the discriminant function derived from the analysis groups. 24 variables were selected to provide a quantitative measure of (1) growth, (2) size, (3) profitability, (4) leverage, (5) dividend policy, (6) liquidity, and (7) stock market characteristics. Of the original 24 variables, seven high market activities, price earnings ratio, past three years' dividends, growth in equity, sales, loss carryover, and the ratio of the last three years' dividends to common equity were found to be significant.

Weston and Mansinghka (1971) carried out an analysis on "Tests of the Efficiency Performance of Conglomerate Companies" to study the performance of the merging companies, pre and post-merger. The study concluded that the rate of earning was low in the control sample group, but after 10 years, performance of both the groups was similar post 10 years of the merger. Achieving growth in the earnings performance of the conglomerate companies was described as a positive outcome of defensive diversification strategy.

"A Multivariate Analysis of Industrial Bond Ratings", research by Pinches and Mingo (1973) applied factor analysis to classify 51 log-transformed financial ratios of 221 companies for four cross-sections six years apart. The study identified seven factors viz., return on investment, capital intensiveness, inventory intensiveness, financial leverage, receivables intensiveness, short-term liquidity, and cash position. These factors explained 78% to 92% (depending on the year) of the total variance of the 51 financial ratios. Moreover, the correlations for the factor loadings and the differential R-factor analysis indicated that the ratio patterns were reasonably stable over time.

Meador, Church and Rayburn (1996) in a study titled "Development of Prediction Models for Horizontal and Vertical Mergers" identified the financial and non-financial variables (market), which predict M&A's of target companies for this active time of business opportunities. A sample of companies that had experienced merger with similar industry and asset size was matched with non-merged companies. The authors used logistic binary regression analysis to identify factors that predicted mergers and acquisitions target companies and then for the subsamples (horizontal and vertical) of the merged companies. The model for horizontal acquisition showed strongest predictive ability with the variables that included long-term debt/total assets, long-term debt/market value, market value/book value, asset growth and sales growth showing significance. It had been earlier suggested that the spate of horizontal mergers and acquisitions during the 1980s was perhaps due to undervaluation of assets; this could, in turn, be traced to high inflation and the conservatism emanating from accounting principles, along with the laissez-faire approach of the Government during the period. This contention was supported by this study.

In an empirical study by Sankar and Rao (1998) titled "Takeovers as a Strategy of Turnaround" the effect of M&A's from the financial angle using certain parameters such as profitability liquidity, leverage and others, has been analysed. They concluded that a sick company can be transformed into a successful one if the acquiring company has better team management skills, and focused attempts are made in the right direction.

A study by Kumar and Bansal (2008) titled "The Impact of Merger and Acquisitions on Corporate Performance in India" analyzed whether Indian M&As able to generate synergy to the extent claimed by the corporate sector. This was done by analyzing the impact of M&A's on the financial performance of the outcomes in the long-run and compared and contrasted the results of merger deals with acquisition deals. The study used ratios and correlation matrix for analysis, and found that in many cases of M&A's, the acquiring companies were able to generate synergy in the long run, which might have been in the form of higher cash flow, more business, diversification, cost cuttings and more.

On the basis of the above literature review the following Hypothesis is formulated:

Hypothesis 1 (H_{1o}): There is no significant difference in the mean score of profitability indicators in the selected units, before and after the merger and acquisition bid.

Harari (1997) concluded that bank merger activity and cost efficiency are positively related, taking into consideration cost efficiency of a bank, economies of scale, and the scope of the Taiwanese banking industry. According to Hopkins (1999), Peng and Wang (2004), Epstein (2005), and Duncan and Mtar (2006) mergers and acquisitions can enhance cost efficiency. Therefore, the pressing query is why some companies perform better than others, post mergers and acquisition. Oftentimes, the answer has been hunted in case studies of the real world.

According to Stahl and Mendenhall (2005) various studies have concluded there that mergers and acquisitions do not generate any additional value for the shareholders, and therefore are considered to be ineffective. They argue that researchers agree that mergers and acquisitions have been taking place time and again in history and will keep on taking place as long as they are alluring and profitable. Therefore, firms opting for mergers and acquisitions must be benefiting from their decision else the popularity of the strategy would drop.

Azhagaiah and Sathish (2011) concluded that M&A deals among the manufacturing companies of India witnessed enhance deficiency of management in the acquiring companies, which was a result of increased profitability, liquidity position, financial, and operating risk and operating performance. Another conclusion of their study was that post merger, the acquiring firms in India are expected to be financially more sound in comparison to the pre-merger period.

Ramaswamy and Waegelein (2003) in an analysis titled "Company Financial Performance Following Mergers" examined the post-merger financial performance of merged companies in Hong Kong to determine relationships between post-merger performance and company size, the compensation plan, method of payment, and industry type, in the long run. The study sample consisted of 162 merging companies from 1975 to 1990, and the analysis covered five years pre and post-merger (using operating cash flow returns on the market value of assets as the measure of performance). The study concluded that there is a positive significant improvement in the post-merger performance; also, there is a significant association between post-merger performance and differences in the relative sizes of the combining companies. It was observed that, firms in different industries i.e conglomerate mergers result in better post-merger financial performance than that of the companies in same or similar industries. It was further reported that larger companies, when acquiring, faced a lot of challenges trying to understand their smaller partners and effectively amalgamate them into the company's workings and culture. Also, corporates with a long term perspective of earnings tend to have a better post-merger financial performance. M&A's from 1983 to 1990 experienced poor post-merger performance in comparison to those before 1983. Therefore, it is an extensive work that not only examined the effect of mergers and acquisitions on long-term performance but determined factors behind such performance.

On the basis of the above literature review the following Hypotheses is formulated:

Hypothesis 2 (H_{2o}): There is no significant difference in the mean score of liquidity indicators in the selected units, before and after the merger and acquisition bid.

Dodd (1980) reported that shareholders of target companies earned large positive abnormal returns from an announcement of merger proposals. These announcement period returns ranged from 13% at the announcement date of the offer to 33.96% average over the duration of the merger proposal, i.e. 10 days before and 10 days after the announcement. On the other hand, shareholders of bidder companies experienced negative abnormal returns of 7.22% and 5.50% over the duration of the proposals.

Asquith (1983) investigated the effect of merger bids on stock returns using the sample of successful and unsuccessful merger bids between July 1962 and December 1976, where the target companies were listed in NYSE. Using the daily common stock returns for two years before the press date until one year after the outcome date, the author concluded that the announcement of a merger bid increases the probability of merger. Further, both successful and unsuccessful target companies exhibited positive and significant average excess returns on the press day and the day before. The examination of the period from the press day to outcome day also suggested that the probability of merger changed during the interim period with new information. The cumulative excess returns rose for successful target companies during this period and fell for unsuccessful target companies. Also, it was concluded that most of the gains from merger go to shareholders of the target companies.

Firth (1990) examined mergers and takeover activity in the UK, specifically focusing on the impact of takeovers on shareholder returns and management benefits. The research shows that mergers and takeovers resulted in benefits to the shareholders of acquired company and to the managers of acquiring company; however, losses were suffered by the shareholders of acquiring company.

Datta, Pinches and Narayana (1992) based on 75 observations for bidders and 79 for targets referred in 41 earlier studies on wealth creation effects of mergers, reported that the bidders, on an average, gained nil or statistically insignificant gains from announcement of mergers while target companies' shareholders experienced over 20% increase in value. The authors further proved that both bidders and targets lost in stock-financed transactions and concluded that of all the factors, mode of payment was the most significant explanatory factor in wealth gains for both bidders and targets. A modest evidence of the positive effect of non-conglomerate mergers on bidder's wealth is also available in this study.

Draper and Paudyal (1999) examined the impact of takeover bid announcement on the returns, trading activities and trading costs of the target and bidding companies. Taking a sample size of 581 target companies and 349 bidding companies between 1988 and 1996, they analyzed daily share prices, volume of trades, number of trades, order size and quoted bid-ask spreads. Based on their studies, the authors concluded that the shareholders of the target companies did benefit from the announcement of takeover bids. Within the 10-day period of the announcement, the cumulative excess returns available to the shareholders of target companies exceeded 11%. In contrast, the shareholders of a bidding company suffered a loss of just under 1% during the same period. Benefits to the shareholders were also found to be dependent on the method of payment. Prices of target (bidding) companies increased (decreased) most if the shareholders of the target companies were given an option to receive the payment in shares or in cash.

Penas and Unal (2004) took a sample of 65 bank mergers and examined the impact of the announcements of a merger on monthly bond returns of acquiring and target-banks. They reported that bondholders of acquiring and target banks realized major positive risk and maturity adjusted bond returns nearing the month of the merger announcement. Also, a major reduction was reported in the credit spreads of the fixed-rate non-convertible bond issues of 38 acquiring banks. Based on cross-sectional regression results, it was clear that after one checked for variables like leverage, diversification, and asset quality changes, the incremental size of the merger product had a significant impact on positive bond returns and reduction in credit spreads. The study also reported that the returns on bond around the time of merger announcement and during the post-merger period declined in spreads were not monotonic with size. Mega-banks and smaller-sized banks did not show any significant announcement month bond returns or post-merger decline in spreads. In contrast, medium-sized banks experienced significant bond returns and realized the reduction in the cost of funds.

On the basis of the above literature review the following Hypotheses is formulated:

Hypothesis 3 (H_{3o}): There is no significant difference in the mean score of leverage indicators in the selected units, before and after merger and acquisition bid.

2.8 Research Gap

India promises further overall growth in various sectors of the economy as it possesses sound economic parameters. It is imperative to have a dynamic environment in order to have positive M&A results. Due to the strength of the Indian economy, there has been huge growth in cross border as well as domestic M&A

activity in the country, in the last decade. The capital market is the mirror showing the strength of an economy. India has emerged as a transparent, mature and a dynamic capital market during the past decade. Thus, India has become a lucrative market for mergers and acquisitions in the country, since the country is on its growing path.

With heightened M&A deals in India coupled with global evidence of M&A improving the shareholder value, it becomes imperative to study the impact of this inorganic strategy on long term performance of Indian corporates. While some studies have already been done in the Indian context, most of them followed a case study approach and lacked specific industry perspective. The present study attempts to give an aggregate picture of the impact of M&A on corporate performance and also examines the impact of M&A deals in the automobile industry.

CHAPTER 3
REGULATORY FRAMEWORK

The present chapter highlights legal provisions relating to takeovers from the USA, UK and Indian perspectives.

3.1 Legal provisions relating to Takeover

USA Perspective

An M&A activity in the USA can be planned through two alternative routes – either a merger route or a tender offer. The acquirer can directly make a financial offer to the shareholders of the target company, based on certain terms and conditions. A tender offer can be negotiated one or it may be a hostile one. US Federal Securities Laws regulate the tender offers, specifying mandatory disclosures and providing substantive regulation. The Corporate Laws of the state where the target company was incorporated also govern the reaction of the target company and their boards of directors to the tender offer, especially the consistency of the target company's actions with its fiduciary duties are checked. Under the circumstance that the offering entity is a controlling shareholder or an affiliate of the target company, the disclosure obligations become stricter. The USA Federal Securities Laws, as well as the concerned State Corporate Laws authorities, would deeply scrutinize the transaction from all aspects. A tender offer structure yields the clear benefit of speediness over other acquisition structures. There are two ways of acquiring control described below.

A. **Privately negotiated transactions:** A bidder can purchase the shares of the target company, privately from one or more shareholders of the target company. Occasionally the target company has an individual or small group of shareholders with the ownership of a controlling interest in it. In such cases, the acquirer needs to purchase the controlling interest from such shareholder(s). Post-negotiation the acquirer may pay the consideration in cash or in any other form and seal the deal of owning a controlling stake in the target company. If a takeover deal is engineered through the direct purchase of controlling interest from a controlling shareholder of the target company, the price offered is usually higher than the market price. The difference between the market price and the offered price has been termed as the 'control premium' and is a measure of the worth of the right to control the target company.

The following approvals are required in this regard:

Stockholders and board of directors: If the acquirer is privately negotiating the purchase of a controlling interest in the target company by interacting with one or more shareholders of the target company, he does not need the approval of the board of directors of the target company or the other shareholders. However, even if it does not require the board's approval, most bidders insist on the same to ensure that the subsequent steps of over-all acquisition go hassle-free.

Hart Scott Rodino (HSR): Sanction under the amended HSR Antitrust Improvements Act, 1976are mandatory even when the shares are being purchased from the target company's shareholders. In case all the requisite criteria are met, the acquirer is supposed to connect with the Federal Trade Commission (FTC) and the Antitrust Division of the United States Department of Justice (DOJ); the purpose is to ensure that there is a detailed analysis of the impact. As per the HSR, there is a mandatory 30days waiting period, which is reduced to 15 days if there is a cash tender offer. During this period, the regulators may require to submit more data pertaining to the transaction or the parties. On several occasions, the timing of the deal is actually determined by the waiting period under HSR.

Securities Law matters: As per an amendment in the Securities Act, 1933 (hereinafter referred to as the 'Securities Act'), the stock sale by the shareholders of the target company has to be mandatorily registered with the Securities and Exchange Commission (SEC). This rule also applies to all other sales or offers of securities. As per the Securities Act, the acquiring company has to file the Schedule 13D (beneficial ownership report) with the SEC, within 10 days of the acquisition of more than 5% of any class of equity security of the company being purchased.

B. **Purchase of shares from the target company:** Instead of acquiring shares from the shareholders, the bidder can opt to purchase it directly from the target company. This can be particularly successful if the target company is in need of funds. If the transaction involves the shift of control from the target company to the bidding company, it becomes imperative for the board of directors of the target company to follow the 'Revlon Duties'. In other words, the directors are responsible for ensuring that the shareholders are making the highest possible short-term gain under the circumstances. For this, the board of directors of the target company must assure themselves that the price they are getting from the bidding company is commensurate with the market rates, and the other bidders, if any, are not offering more profitable than the deal. If the target company will lose control as per the bidding transaction, then the acquirer is likely to face one of the following two propositions from the target company's board of directors:

i. Ensure payment of the control premium, or

ii. Ensure some mechanism of a 'standstill' arrangement that prevents the acquirer from purchasing more shares or exercising control over the target.

The following approvals are required in this regard:

Shareholders: In case the number of shares that needs to be issued to give the bidder a controlling interest in the target company is very large, and more than the number of authorized but unissued shares of the target company, then this will require the amendment of the "Certificate of Incorporation" (COI) of the target company to enhance the number of its authorized shares. This shall entail the approval of the shareholders of the target company. For this approval, the target company needs to organize a shareholder's meeting and seek proxies for votes. The target company needs to prepare a proxy statement and get it cleared by the SEC. Also, in accordance with the Securities Exchange Rules, the shareholders need to agree to significant issuances.

Board of directors: The board of directors of the target company needs to agree to the issuance or amendment, if required, of the 'Certificate of Incorporation' HSR approval is also needed to allow the purchase of the newly-issued shares, if all requirements are met.

Securities Law matters: Issuing new shares by the target company requires its registration with the SEC. Alternatively, the target company must qualify for an exemption from this requirement. In addition, after the purchase, the acquirer will need to file a Schedule 13D with respect to its interest in the target company.

Indemnification: The acquiring company may insist to be indemnified by the target company in case there have been misrepresentations regarding the target company's assets or businesses in the share-purchase agreement. The target company's representations regarding its own business details are more likely to be accurate than in deals involving the direct purchase of shares from shareholders.

Deal certainty: The board of directors of the target company would aim at securing the right to back out of the transaction or at least modify its recommendations if there is the chance of receiving a better offer in the course of time.

Consequences of issuance: As per the COI's of some target companies, the shareholders may possess pre-emptive rights on new issuances. This may prevent the shareholders from giving the company's controlling interest to a potential bidder. Also, options, warrants or securities convertible into shares of the target company may have anti-dilution clauses, which will be affected by an issuance.

Acquisition through merger transactions: In case of a single-step merger, the acquirer and the target company will enter into a merger agreement through negotiation. The approval of the board of directors and the shareholders of the target company (under the relevant State Corporation's law statute) is required for the agreement of merger. In the state of Delaware, USA, for example, the normally required approval is a majority of the outstanding shares. The transaction is deemed complete once approval by the shareholders has been received and all other formalities of the completion of the merger have been satisfied.

C **Disclosure requirements:** According to the proxy rules, the onus of preparing and filing the proxy statement with the SEC lies on the target company and not on the purchasing company. Post the approval from the SEC, the target company is required to share the same with its stockholders for them to approve the single-step merger. In case there are major changes in the proxy statement once it has been shared with the stockholders, it is again the responsibility of the target company to send it as an addendum or corrigendum or supplemental document to its shareholders.

Few major items that must be disclosed are:

i. Summary of the material terms of the transaction;
ii. Details of earlier deals and main corporate events between the bidder and the company to be purchased;
iii. Background of the deal, detailing the history of the deal negotiation;
iv. A brief summing up of the various points explaining why the board of directors of the target company are in favor of being acquired by the bidder;
v. Details of document, opinion or appraisal inherently related to the bid, which has been shared with the board of directors of the company being purchased; and
vi. Financial details for the bidder, if financing for the deal is unsure.

UK Perspective

'The Blue Book' contains all the rules pertaining to a takeover deal (public companies only) in the UK, officially controlled by the 'City Code' on takeovers and mergers. Initially, the Code used to be a non-mandatory set of rules with city authorities considering its criteria voluntarily. However, in 2006 the guidelines were made mandatory, including it as a part of the UK's compliance with the European Takeover Directive.

As per this code, each shareholder of a company is equal to the other. It controls the information that may or may not be openly shared when a takeover bid is underway. The timescale of certain bid-related events is also set by the Code. Most importantly, the Code sets the least bid price levels based on previous procurement of shares.

M&A in UK Company Law

The Company Law in the UK regulates M&A (including, reconstructions or takeovers) for general reconstructions, amalgamations, demergers, and takeovers, which concern acquisitions of public companies.

Established in 1968, the 'Panel on Takeovers and Mergers' regulates for the provisions of the Companies Act, as well as those of the European Directive on takeover Bids (2004/25/EC) for public companies.

The Panel on Takeovers and Mergers

The Panel on Takeovers and Mergers (the "Panel") is the body that regulates takeovers of companies that are subject to the City Code.

European Commission

The European Commission has absolute authority to decide on issues resulting from proposed takeovers, in certain cases, since the United Kingdom is a member of the European Union (EU).

The City Code

As per the "Directive on Takeover Bids", the Panel has clear regulatory powers to govern merger and takeover bids. The City Code has been prepared and conducted by the Panel. The City Code has jurisdiction in the United Kingdom and the Panel can exercise its powers in all the deals and transactions to which the City Code is applicable.

With 6 general principles, 38 rules and several notes to help interpret the rules, the City Code is diverse and rich. All the 38 rules of the City Code are expanded versions of the general principles and provide regulations to control each aspect of a takeover bid. Needless to say, these rules need to be followed both in the word and in spirit.

Importantly, the profit or loss made either by the target company or the bidding company is not in the purview of the Panel. It is not the prerogative of the City Code to either facilitate or to obstruct the takeover offers made.

The chief objectives of the City Code can be condensed into the following 3 pointers:

i. the target company's shareholders of the same class must be meted out equal treatment with respect to access to relevant information. None of the shareholders should complain of a lack of knowledge that subsequently affected their decision-making.

ii. no action of the target company's top management should deter the bidder's offer without its shareholders' agreement.

iii. under no circumstances, there should be the creation of a fake market for the securities of the bidder of the target company.

The code ensures that the treatment meted out to each shareholder is the same, and none receives preferential consideration. The code also controls the information pertaining to a bid that companies are supposed to hold onto or release to the public and sets the timeline for certain actionable items of the bid. Most importantly, it finalizes the minimum bid levels as per the earlier share purchases.

Important features contained in City code are:

i. the shareholder's offer must come when his or her stock, inclusive of PAC, is 30% of the target. This is the mandatory bid rule;

ii. there should not be any leaking out of the bid-related information except through regulated announcements;

iii. the onus is on the acquiring company to make a formal announcement;

iv. the offer must not be lower than the price paid by the acquirer within a period of 3months prior to announcing the offer;

v. if shares are purchased within the offer period at a price which is more than the offer price, the offer must be increased accordingly.

Indian Perspective

The law relating to takeovers in India is contained in the Companies Act, 2013 and the Securities and Exchange Board of India (Substantial Acquisition of Shares and Takeovers) Regulations, 2011 (SEBI Takeover Code).The takeover of companies listed on at least one of the recognized Stock Exchanges in India is regulated by the provisions of the Listing Agreements of various Stock Exchanges.

Company Law

Share acquisition is permitted after obtaining the approval of the board of directors as well as the audit committee of the acquiring company. Approval of the shareholder is also necessary for the sale of shares between related parties.

As per the procedure, a takeover proposal has to be first approved by the audit committee, followed by the board of directors; if the company is listed in the stock exchange, then the stock exchange is next in line. The approval has to be obtained from the shareholders with the required level of majority i.e. a majority in the number of shareholders and 75% in value of shareholders/creditors through personally voting, using postal ballot/proxy. The scheme of the amalgamation as approved by the shareholders of the company has to be submitted to the National Company Law Tribunal (NCLT), the Court, for its approval. The Court must give notice for every application made to it for this purpose, to the Central Government.

NCLTis the final approving authority who will approve the same after considering the feedback of significant authorities including the Regional Director, Official Liquidator, Registrar of Companies, RBI, SEBI, Stock Exchange, Income Tax Authorities and Competition Commission of India (CCI). Before approving the deal, the NCLT is also supposed to go through objections, if any, from concerned stakeholder(s).

Income-tax Act, 1961 (ITA)

Capital Gains tax, in case of a slump sale/sale of shares of a company, is chargeable on the net capital gains depending upon the time period for which the undertaking/shares are held. In case of a sale of a listed company's shares, the capital gains arising on transfer of such shares on the stock exchanges would be exempt from capital gains tax or would be chargeable at the reduced rates depending on the period for which such listed shares were held. A classical amalgamation or demerger, involving a continuation of at least 3/4th shares (in the value of shares of the transferor company) to shareholders, is a tax-neutral transaction under ITA, subject to the satisfaction of other specified conditions. ITA also provides for carrying forward of business losses in the transferee entity subject to fulfillment of certain other conditions.

Securities Laws

If more than 25% of shares of a listed company are acquired, it shall activate an open offer to the public shareholders. However, as per the Takeover Code, in a listed company's merger or demerger open offer to the public shareholders is generally not initiated. In case of a merger or demerger relating to a listed company, approval in advance by the stock exchanges and SEBI be obtained, even before NCLT has been approached.

In the takeover process, certain legal terminologies are commonly used, which have been defined as below:

Acquirer: An acquirer is an individual, legal entity or a corporate company that either by itself or with Persons Acting in Concert (PAC's) acquires voting rights or shares or control over a target company, either directly or indirectly.

Persons Acting in Concert (PAC's): Persons acting in concert, abbreviated as PAC's, are natural persons or business entities or other legal entities who come together with the common purpose of acquiring shares or voting rights in or gain control over a target company. Their mutual agreement can be formal or informal and their cooperation can be direct or indirect.

Target Company: A company or corporation with public-listed equity shares that attracts the bidding interest of a potential acquirer is called the target company.

Control: The term "Control" implies the power to exercise certain rights with respect to the target company. These rights include the right to appoint a majority of the directors, the right to managerial control and the right to policy-related decision making. These rights can rest with any individual or PAC's or corporate or legal entities by way of shareholding, voting agreements, shareholders' agreements or management rights or through any other legal way. These rights can be exercised either directly or indirectly.

Various provisions relating to the acquisition of shares of a company are discussed below:

Takeover Code

"Securities and Exchange Board of India" (**SEBI**) is the main authority with the aim of regulating those entities that are either already listed or are about to be listed on the Stock Exchanges in India. The Securities and Exchange Board of India (Substantial Acquisition of Shares and Takeovers) Regulations, 2011 (the "Takeover Code") restricts and regulates the acquisition of shares, voting rights and control in listed companies. Acquisition of shares or voting rights of a listed company, entitling the acquirer to exercise 25% or more of the voting rights in the target company or acquisition of control obligates the acquirer to make an offer to the remaining shareholders of the target company. Takeover Code provisions become applicable if a takeover bid is being planned through the issuance of new shares, or acquisitions of voting rights or existing shares of a listed company.

The major aim of the Takeover Code is to ensure that the shareholders of a listed company have access to all the relevant information regarding an imminent change in the ownership of a company and to provide them with an exit route (though restricted) if they want one. As per code, there can be various ways of takeovers described herein below:

i) Mandatory offer: As per the Takeover Code, the acquiring company must give the other shareholders an exit option while making them the offer to purchase their shares. This obligation to make a compulsory open offer becomes active under the following circumstances:

   a) Initial trigger: Where the acquisition of shares or voting rights in a target company entitles the acquiring company along with the PAC to exercise 25% or more of the voting rights in the target company.

   b) Creeping acquisition: If the acquirer already holds 25% or more and less than 75% of the shares or voting rights in the target, then any acquisition of additional shares or voting rights that entitles the acquirer along with PAC to exercise more than 5% of the voting rights in the target company in any financial year.

   c) Acquisition of control: If the acquirer acquires control over the target, regardless of the level of shareholding, acquisition of control of a target company is not permitted without complying with the mandatory offer obligations under the Takeover Code. The definition of 'control' is usually subjective and varies from one case to another. The Takeover Code has defined control as to include: right to appoint majority of the directors; right to control the management or policy decisions exercisable by a person or PAC, whether directly or indirectly, due to their stockholding or management rights or voting agreements or shareholder agreement or in through any other valid mechanisms.

ii) Indirect acquisition of shares or voting rights: As per the Takeover Code, there are clear guidelines to manage indirect acquisitions. As per the guidelines, "any acquisition of shares or control over a company, business or entity that would enable a person and persons acting in concert with him to exercise such percentage of voting rights or control over the target company, which if directly acquired in the target company would have otherwise necessitated a public announcement for open offer, shall be considered an indirect acquisition of voting rights or control of the target company".

iii) Voluntary open offer: A shareholder, who lays claim to 25-75% of the shares or voting rights in a company, is allowed to announce an open offer for purchasing more shares. However, the total shareholding after the open offer is closed, should not be more than 75% as this would be a breach of the maximum non-public shareholding limit. The minimum offer is at least 10% of the shares. As per SEBI's Takeover Code, any shareholder owning less than 25% shareholding or voting rights is also eligible to make a voluntary open offer for purchasing additional shares. Regulation 8 of the Takeover Code outlines the criteria to calculate the offer price to be paid to the public shareholders. This holds true in case of a voluntary open offer and the mandatory open offer. A few additional guidelines have been prescribed for ascertaining the price when the open offer is made in accordance with an indirect acquisition. Importantly, an acquirer cannot decrease the offer price but an upward correction of the offer price is allowed under special conditions.

iv) Competitive Bid/Revision of Offer/Bid: The Takeover Code gives an option to a person, by making a public announcement, to purchase the shares of the target company through a competitive bid. Such a person is known as the first bidder, not the acquirer. However, a period of not more than 15 days from the date of the detailed public announcement of the first bidder is given to make the bid. The competitive bid must be for at least the number of shares held or agreed to be acquired by the first bidder (along with PAC), plus the number of shares that the first bidder had bid for.

Listing Regulations

The listing regulations of the Stock Exchange have outlined a very detailed framework to control various types of listed securities. Regulations 30 of the listing regulations cover the disclosure of material events by the listed entity. Such an entity must clearly disclose events specified under Part A of Schedule III of the listing regulations. The listing regulations divide the events that need to be disclosed broadly into two categories. The events that have to be necessarily disclosed without applying any test of materiality are indicated in Part A of Schedule III of the listing regulations. Part B of Schedule III includes the events that ought to be disclosed by the listed entity, if considered material.

Foreign Exchange Regulations

According to the guidelines for pricing and permitted sector wise caps, set by the RBI, residents and non residents are allowed to make transfer of or sell the equity shares. In case of a typical merger/demerger, where shares are issued to a non resident shareholder of the company being transferred, a pre sanction of the RBI or the government is not required with the condition that the transferee company stays within the limits of the sector wise cap set for foreign exchange and the courts in India have given an approval for the merger or demerger. A pre approval by the RBI would be needed in case of issuance of any instrument apart from equity shares/ compulsorily convertible preference shares / compulsorily convertible debentures to the non residents since they are treated as debt.

Based upon the above discussion it can be said that rules and regulations have been made over the years relating to mergers and acquisitions. The legal &regulatory legal aspect of M&A broadly includes the following:

Proposal analysis: The management of the companies that are planning to merge or amalgamate should analyze the pros and cons of the deal. The future prospects of growth of the target company must be evaluated in comparison to the actual contribution of the firm after the deal of acquisition.

Negotiation of exchange ratio: The merger or acquisition happens through the exchange of shares for which exchange ratio is to be negotiated.

Approval of the board of directors: The board of directors of both the companies has to approve the final scheme of acquisition that evolves as a result of negotiations.

Approval of Shareholders: The scheme as approved by the respective boards is placed before the shareholders of the respective companies for their approval.

Consideration of interest of the Creditors: The scheme should also be approved by the majority of creditors in number and $3/4^{th}$ in value so that their interests are protected.

Approval by National Company Law Tribunal (NCLT), the Court: The Court is the final approving authority. The NCLT would approve the scheme only when it is satisfied that the scheme is just, equitable and reasonable for all concerned and it is not against the public interest.

CHAPTER 4
RESEARCH METHODOLOGY

The present chapter explains the significance, objectives of the study followed by hypotheses formulation, and research design. This study aims to examine the impact of corporate takeovers on the growth & financial performance of selected companies under automotive industry, to investigate the growth drivers and challenges for automotive industry, to identify the future trend of automotive industry, in India, and to give recommendations for improvement in growth and development of the automobile industry. This involved an exhaustive study of the takeover practices being followed in India, UK and US and their impact on the growth and financial performance of the automobile industry. The study is descriptive and empirical in nature. To achieve the objectives of the study, primary as well as secondary data has been used.

4.1 Significance of the study

This chapter of the study indicates that M&A is a value-maximizing decision, driven to maximize the company's value. M&A are effective strategies and help the acquirer companies in attaining their underlying intents. M&A more likely seems to be the strategies used by corporate companies for the attainment of manifold objectives. Cost economies, as expected due to large scale or resource sharing; perhaps seem to be a distant possibility in present competitive environment. In this view, attaining cost competitiveness and sustainable growth, meeting market challenges better (domestic as well as international), acquiring brand-name, collusive synergies by merging with competitors, portfolio enrichment, ease to market entry, enhancing customer base, etc. seem to be more practical rationales for corporate companies.

The real push for takeovers in India came in 1991 after the economic reforms were introduced by the then Government. The economic reforms focused on increasing competition, improving efficiency, through loosening controls and regulations on production, trade, and investment. With new economic reforms also came the challenge of competition. Indian companies faced tremendous pressure from local as well as global players. To meet these challenges, it became imperative to increase efficiency and reduce costs. In the face of this, Indian companies looked into to grow inorganically, through expansion and by way of mergers and acquisitions. Further, positive regulatory changes facilitated such moves of Indian companies. For instance, the amendment of the MRTP Act made it possible for group companies to consolidate through mergers, eliminating duplication of resources and bringing down the costs. Post economic reforms takeover became a viable strategy for growth in India due to the easing of regulation, restructuring of family-owned conglomerates, sale of state-owned companies, overcapacity, and deregulation of fragmented industries.

4.2 Objectives of the Study

This study aims to analyze the impact of the corporate takeover on the automotive industry. Based on the research gap, the following objectives have been set:

i. To review the past and present scenario of merger and takeover practices in US, UK, and in India

ii. To examine the impact of the corporate takeover on the growth and financial performance of the selected companies under the automotive industry.

iii. To investigate the growth drivers and challenges for the automotive industry.

iv. To identify the future trends of the automotive industry, in India.

4.3 Hypothesis

Based on the literature review, and experience survey the following hypotheses are formulated under the study:

$H1_o$ There is no significant difference in the mean score of profitability indicators in the selected units, before and after the merger and acquisition bid.

$H2_o$ There is no significant difference in the mean score of liquidity indicators in the selected units, before and after the merger and acquisition bid.

$H3_o$ There is no significant difference in the mean score of leverage indicators in the selected units, before and after the merger and acquisition bid.

4.4 Research Design

The ex post facto research design has been employed in this study. The rationale behind selecting this design is the fact that it is used where the phenomenon being investigated has already occurred. Hence, the approach is to collect primary as well as secondary data and analyze it to ascertain the relationship between the parameters being studied. After the review of existing literature and from the collected data, research has been designed to meet the objectives of the study.

4.5 Data Collection, Sample Size, and Sampling Techniques

The primary data was collected to represent facts using the method of questionnaire formulation after doing informal discussion to understand the concept. The variables were specifically structured keeping in view the objectives of the present research, to study the perception of professionals from the connected field about the value of the company, growth drivers, managing growth during the transition, future plan and stakeholders interest in successful corporate takeovers in the automobile industry.

The sample was identified using non-probability purposive sampling technique of 250 professionals. For the purpose of this study, a dispersed representation of demographics like gender, profession, the experience was taken. Finally, the data has been collected and analyzed based on responses received from 211 respondents. The secondary data has been collected from the published reports of the selected companies. Other information related to these companies has been collected from the official website and web sources, annual report, books, publications, and journals. Multiple sources of data have been used since there is no single official database on M&A in India which gives a complete picture of M&A. For the purpose of examining the impact of M&A on long term operating and financial performance of companies, accounting ratios have been used as data variables along with financial variables.

A fundamental selection criterion to select companies was that selected companies retain their corporate identities before and after the M&A deal. Also, the selection criteria of the companies selected for study includes:

1. Their products

2. Their Brand image

3. Stock market indices i.e. stock price movement over 5 years

4. Their operation and financial strengths.

For this purpose the following four companies have been selected as the sample for the study:

S. No.	Name of the Company
1	Tata Motors Ltd.
2	Volkswagen
3	Mahindra & Mahindra Ltd.
4	Daimler Benz

The objectives and motives behind the acquisition by the aforesaid companies are discussed hereunder:

Acquisition of Jaguar Land Rover (JLL) by TATA Motors Ltd. in 2008

Tata Motors bought off Jaguar and Land Rover (JLR) from the US automobile giant Ford Motors. This purchase was inclusive of IPRs, JLR's manufacturing units, its UK based twin advanced design centers, and its national sales companies spread across the world. Markets reacted with certain incredulity at this development, considering that it was the first time an Indian company had pulled through an expensive deal, acquiring a luxury automobile brand. Tata Motors made all-round gains with this acquisition of JLR. First and foremost, Tata Motors got an express entry ticket into an elite segment of the world automobile market, becoming a global presence to be reckoned with, almost overnight.

Besides, Tata Motors access to JLR's advanced design studios and its sophisticated technology is bound to improve its product quality in India. Further, this deal propelled Tata Motors onto the global platform with the mere stroke of a pen – a feat that would have otherwise taken years of toil. Tata Group also gained the advantage of associating with Corus, who supply automotive high-grade steel to JLR and other globally significant automobile manufacturers.

Lastly, Tata Motors is assured to spread its market reach beyond India and South East Asia, which are presently contributing to the major chunk of its sales. In reciprocation, Jaguar Land Rover is bound to gain by entering Tata's stronghold of SE Asia, reducing its dependence on American and European markets (30% and 55%, respectively).

Acquisition of Porsche by Volkswagen (V.W) in 2012

The Germany-based automotive giant Porsche is a specialist in sports cars and a new line of all terrain vehicles. With models such as the Boxster and Cayenne and impeccable quality assurance, Porsche enjoyed an iconic status across the globe as well as financial gains during the mid-2000s. Its low manufacturing depth is the key element, making Porsche's business model unique and profitable.

Volkswagen AG, too, is a key member of the automotive industry, recognized as a manufacturer of passenger and commercial vehicles. Audi, Scania, SEAT, Skoda, and Bentley are the brands operating under the V.W group. Such an impressive line-up of brands in its kitty enables V.W to gain a substantial edge over its competitors. In fact, the year 2007 saw the V.W group sell â,¬6.2 million vehicles, enjoying a 9.8% share in the global passenger car market. However, the rising material prices were a major threat to V.W.

Volkswagen aimed at the merger as a protective stance against a hostile takeover. It also hoped for an up thrust from Porsche's positive image and an applauded managerial team. V.W was also hoping for a financial gain from this merger. Despite 15 times higher annual revenue than Porsche, V.W had seven times lesser profit margins than Porsche. V.W profit margins were less than 1%, with profits of â,¬484 million on sales of â,¬55.4 billion in the first half of calendar 2005.

Even for Porsche, this deal was magnetically attractive, with their attractive and popular vehicle designs being co-produced by V.W., where V.W had a wide range of products and Porsches focus was on luxury products. Thus, this merger clearly provided diversification benefits to the latter. Rising material costs were also a major threat to Porsche. Under these circumstances, considering the benefits to both, V.W bought 50.1% stake in Porsche. Balance 49.9% stake was already with V.W.

Acquisition of Ssang Yong by Mahindra & Mahindra Ltd. (M&M) in 2010

The acquisition was significant for M&M, as it saw the chances for a prosperous partnership between Ssang Yong and M&M when not only the quality would be consumed but the brand identities would be protected as well. A council known as synergy council comprising the officers from the two companies was created which would ensure focus and delivery of synergies between both the companies. A number of aspects such as new car development, global procurement and business strategy were the main focus of the council so that international markets could be penetrated. The launch of Rextonand Korando-C in India is a part of the strategic plan of the India project, which has already taken off. There were a lot of discussions relating to the bright chances for joint production and technological developments and synchronization of worldwide procurements and operations. Ssang Yong has been reviewing for its suitability, the solid IT system of M&M.. The company has also been planning to set up the operation of Mahindra Finance Ltd. in Korea so that the sales of Ssang Yong vehicles can be improved.

Ssang Yong carries with itself a wealthy heritage of R&D and innovation, since it is an eminent automotive company. This heritage, combined with the synergies between Ssang Yong and M&M will prove to be a super power in the global utility vehicle space because of the combined R&D, product development and platform sharing between the two companies.

M&M has also proposed the following five point agenda for Ssang Yong:

i. Strengthening the product pipeline;
ii. Harnessing synergies between the two companies;
iii. Investing in the brand;
iv. Building human resources;
v. Focusing on financial stability;

Ssang Yong has also proposed the following investments:

i. In 2011, the business plan calls for a 70%investment increase in product development, as compared to last year, at over KRW 200 billion

ii. Over 40 billion KRW for brand building in Korea - a 60% increase over 2010, and an increase in overseas brand investment by over four times in 2011.

Under these circumstances, considering the benefits to both, M&M acquired Ssang Yong

Acquisition of Chrysler by Daimler Benz in 1998

The acquisition of Chrysler by Daimler Benz was deemed a failure. The deal appeared to be mutually beneficial as Daimler Benz could now enter the American market while Chrysler, with a flattened growth curve within the USA, gained access to global markets. This deal had envisaged several mutually beneficial achievements, it was expected that German luxury auto-making technology would be united with the fast-paced mass marketing skills of American businesses.

The Daimler-Chrysler joint entity aimed at overtaking Japanese, German and American automotive giants through reduced expenses and benefits of economies of scale. The high manufacturing costs of Daimler was expected to be inspired by the low-cost development and efficient entrepreneurial culture of Chrysler. The comprehensive distribution network of Daimler spread across the globe was expected to benefit Chrysler.

However, this merger failed to meet expectations. The market share of Chrysler in the USA dipped to 13.2% (2003) from its 1996 level of 16.2%.

An analysis revealed that Chrysler had failed to offer product diversity in the face of stiff competition in the segment of SUVs and mini-vans. Given the aged old model line-up, a fall in sales was inevitable. With the Euro strengthening as compared to the US Dollar (in 2003), Chrysler experienced a slide in revenues; low sales and high sales incentives contributed to the loss. The senior management in Chrysler saw a major upheaval, following the merger, some retired, a few quits while the others were simply discharged. This took away the very essence of efficient entrepreneurship that the merger planners were hoping, would be imbibed by the Chrysler-Daimler chimera.

The Chrysler-Daimler hybrid lost market share globally, reducing to 7.97% in 2003. Merger hiccups and plummeting sales at Daimler were cited as the causes.

4.6 Statistical Tools and Techniques for Analysis

After the data is collected, it was coded and subsequently transferred to the statistical package for Social Sciences (SPSS) software for analysis. Before the analysis, the data was passed through the cleaning process, a way to check the mistakes in the entered data. Cleaning process involved count, frequency distribution and average computations.

Preliminary data analysis included relative frequency distribution, descriptive analysis to measure the central tendency and dispersion of variables. The descriptive used to summarize the data in this study are the mean, maximum, minimum and standard deviation. All statistics are properly tabulated and presented. Every table has a clear and concise title to make it understandable.

The study used Cronbach's alpha coefficient to examine the internal consistency (reliability) of the items used in the questionnaire.

After validating the instrument using Cronbach's coefficient, exploratory factor analysis (EFA) is conducted to explore the factor loading on the latent variables and their variances and co-variances in the observed indicators. EFA is useful for identifying the underlying factor structure and providing convergent and discriminant validity, it assumes that the measurement errors of the items are correlated. The study used Kaiser-Meyer-Oklin (KMO), a measure of sample adequacy for all latent variables to validate the factor analysis.

The t-test is utilized in the study to test the null hypothesis that the means of two groups are equal. The study used specifically independent sample t-test and was computed using SPSS statistical program. In fact, as the sample size in the present study is fairly large, the t-test is assumed to be valid (i.e. the type I error rate is controlled at 5%) even when X doesn't follow a normal distribution. Because of the central limit theorem, the distribution of these, in repeated sampling, converges to a normal distribution, irrespective of the distribution of X in the population.

The F Test is utilized to test the null hypotheses that the mean of groups does not differ significantly. If p is less than a pre-determined threshold (for example, $\alpha = 0.05$)

the null hypotheses are rejected at 5% level of significance and the factors are deemed to have a significant effect.

For the purpose of examining the impact of M&A on long term operating and financial performance of acquirers, various accounting ratios have been used as data variables along with financial variables.

CHAPTER 5
DATA ANALYSIS, RESULTS AND DISCUSSIONS

The present chapter consists of results and discussion on the basis of data analysis and existing literature. This chapter includes the findings of the study based on primary and secondary data analysis.

5.1 Analysis of Primary Data

A detailed analysis of the collected data was conducted and is reported. The data were processed and analyzed with the Statistical Package for the Social Sciences (SPSS) software 19.0, using statistical tools of frequencies, descriptive statistics, Cronbach's alpha–reliability test, statistical tests of hypothesis t and F, KMO & Bartlett's test, Factor analysis.

Frequencies & Percentages

Frequencies: Frequency distribution of the variables provides a glimpse of the data collected in the primary survey. A frequency is the number of occurrences in a single variable. In other words, it is a count of individual responses. The information that is received in a frequency can serve as the foundation for other calculations.

Percentages: Percentages explain information as a proportion of the whole. Percentages are calculated by taking the number of the subcategory and dividing by the total number in the population. Percentages are useful for understanding group characteristics or group comparisons.

Tables 5.1 indicate the demographic profile of respondents according to their gender, profession, experience, and age. Total 211 respondents have given their views.

Table 5.1: Demographic profile of respondents based on their gender, profession, experience, and age

Gender	Profession	Profession No.	Profession %	Experience No. of years	Experience No.	Experience %	Age	Age No.	Age %
Male	Chartered Accountant	93	62.4	<1 year	19	12.8	<25 Years	21	14.1
	Company Secretary	34	22.8	1-5 Years	21	14.1	25-35 years	36	24.2
	Cost Accountant	3	2.0	6-10 Years	16	10.7	35-45 years	32	21.5
	Advocate	0	0.0	11-20 Years	31	20.8	> 45 years	60	40.3
	MBA	5	3.4	>20 Years	62	41.6			
	Other	14	9.4						
	Total	149	100		149	100		149	100
Female	Chartered Accountant	18	29.0	<1 year	14	22.6	<25 Years	21	33.9
	Company Secretary	31	50.0	1-5 Years	25	40.3	25-35 years	30	48.4
	Cost Accountant	1	1.6	6-10 Years	12	19.4	35-45 years	8	12.9
	Advocate	1	1.6	11-20 Years	8	12.9	> 45 years	3	4.8
	MBA	1	1.6	>20 Years	3	4.8			
	Other	10	16.1						
	Total	62	100		62	100		62	100
Total	Chartered Accountant	111	52.6	<1 year	5	2.4	<25 Years	42	19.9
	Company Secretary	65	30.8	1-5 Years	11	5.2	25-35 years	66	31.3
	Cost Accountant	4	1.9	6-10 Years	14	6.6	35-45 years	40	19.0
	Advocate	1	0.5	11-20 Years	59	28.0	> 45 years	63	29.9
	MBA	6	2.8	>20 Years	122	57.8			
	Other	24	11.4						
	Total	211	100		211	100		211	100

Source: SPSS output from field survey data

Tables 5.1 indicate the demographic profile of respondents according to their gender, profession, experience, and age. Out of total 211 respondents, 149 respondents are male and 62 are female, 111 are C.A (Chartered Accountant), 65 are C.S (Company Secretary), 4 are Cost Accountant, 1 is Advocate, 6 are MBA, and 24 are other, respectively, Also Out of 211 respondents, 5 respondents are having experience less than 1 year, 11 are having experience 1-5 years, 14 are having experience 6-10 years, 59 are having experience 11-20 years, and 122 are having experience more than 20 years, respectively. Further, Out of 211 respondents, 42 respondents are aged below 25 years, 66 respondents are aged between 25-35 years, 40 respondents are aged between 35-45 years, and 63 respondents are aged above 45 years, respectively.

Table 5.2 indicate the respondents views regarding effect on the value, growth drivers, future trend and reasons with respect to mergers and acquisitions of companies

Table 5.2: Respondents views regarding effect on the value, growth drivers, future trend and reasons with respect to mergers and acquisitions of companies

Item	Statement	Strongly Disagree	Disagree	Neutral	Agree	Strongly Agree	Total Agree
A_1	The disclosure of return on equity affects the value of a Company.	2.4%	5.2%	6.6%	28.0%	57.8%	85.8%
A_2	The takeover substantially affects the performance of a Company	0.0%	0.0%	7.6%	36.0%	56.4%	92.4%
A_3	The corporate takeover affects the interest of the stakeholders	0.0%	0.9%	3.8%	39.3%	55.9%	95.3%
A_4	Geographic coverage are the Company's primary reason for an acquisition	0.0%	1.9%	7.6%	45.0%	45.5%	90.5%
A_5	The growth driver in takeover of a Company is operating synergy	0.0%	0.0%	11.4%	42.2%	46.4%	88.6%
A_6	The sharing of information about the future growth potentials plays an important role in the success of corporate takeover	0.0%	2.4%	11.4%	41.7%	44.5%	86.3%
A_7	The primary reasons of failure of a corporate takeover include corporate cultural differences.	0.0%	4.3%	9.5%	43.1%	43.1%	86.3%
A_8	Proper communication among the stakeholders during the transition period is necessary to mitigate the failure of a corporate takeover.	0.0%	0.0%	6.2%	42.2%	51.7%	93.8%
A_9	Measuring market trends for future growth potential of Company are important for success of corporate takeover.	0.0%	0.5%	8.5%	41.2%	49.8%	91.0%
A_{10}	In case of takeover of a Company initially the productivity may drop temporarily as people take time to become familiar with new systems	0.5%	6.6%	9.0%	34.1%	49.8%	83.9%
A_{11}	India is an anchor of future growth in the auto industry	0.5%	1.4%	11.4%	36.0%	50.7%	86.7%
A_{12}	The most popular form of vehicle in India are 2 wheelers / passenger cars	0.0%	1.9%	9.5%	35.5%	53.1%	88.6%
A_{13}	The factors that contribute to an increase in India's automobile demand is increasing buying power of customers	0.0%	0.5%	7.1%	38.9%	53.6%	92.4%

Item	Statement	Strongly Disagree	Disagree	Neutral	Agree	Strongly Agree	Total Agree
A_{14}	The automobile industry of tomorrow will reflect increased prosperity	1.4%	4.3%	11.4%	36.0%	46.9%	82.9%
A_{15}	The main challenges in the growth of Indian automobile industry includes fluctuations in fuel prices	1.4%	3.3%	7.6%	35.5%	52.1%	87.7%
A_{16}	Availability of skilled labor at low cost is important growth driver of automobile industry	0.0%	2.4%	17.1%	20.9%	59.7%	80.6%
A_{17}	Robust Research and Development centers affects the growth of automobile Industry	0.0%	0.0%	4.7%	40.8%	54.5%	95.3%
A_{18}	Greater fuel efficiency is important growth driver of automobile industry	0.0%	0.5%	15.6%	31.3%	52.6%	83.9%
A_{19}	New segment of customers are growth driver of automobile industry	0.0%	0.5%	11.8%	52.1%	35.5%	87.7%
	Average						88.4%

Source: SPSS output from field survey data

Table 5.2 shows the respondents views regarding the effect on the value, growth drivers, future trend and reasons with respect to mergers and acquisitions of companies. It indicates that **88.4%** of respondents agree with the corresponding statements.

Table 5.3 shows the descriptive statistics of the views of respondents on the corresponding statements.

Table 5.3: Descriptive Statistics

Item	Statement	Mean	Standard Deviation	Minimum	Maximum
A_1	The disclosure of return on equity affects the value of a Company.	4.34	0.98	1.00	5.00
A_2	The takeover substantially affects the performance of a Company	4.49	0.64	3.00	5.00
A_3	The corporate takeover affects the interest of the stakeholders	4.50	0.62	2.00	5.00
A_4	Geographic coverage are the Company's primary reason for an acquisition	4.34	0.70	2.00	5.00
A_5	The growth driver in takeover of a Company is operating synergy	4.35	0.68	3.00	5.00
A_6	The sharing of information about the future growth potentials plays an important role in the success of corporate takeover	4.28	0.76	2.00	5.00
A_7	The primary reason of failure of a corporate takeover includes corporate cultural differences.	4.25	0.80	2.00	5.00
A_8	Proper communication among the stakeholders during the transition period is necessary to mitigate the failure of a corporate takeover.	4.45	0.61	3.00	5.00
A_9	Measuring market trends for future growth potential of Company are important for success of corporate takeover.	4.40	0.66	2.00	5.00
A_{10}	In case of takeover of a Company initially the productivity may drop temporarily as people take time to become familiar with new systems	4.26	0.91	1.00	5.00
A_{11}	India is an anchor of future growth in the auto industry	4.35	0.77	1.00	5.00
A_{12}	The most popular form of vehicle in India are 2 wheelers / passenger cars	4.40	0.74	2.00	5.00
A_{13}	The factors that contribute to an increase in India's automobile demand is increasing buying power of customers	4.45	0.65	2.00	5.00
A_{14}	The automobile industry of tomorrow will reflect increased prosperity	4.23	0.91	1.00	5.00
A_{15}	The main challenges in the growth of Indian automobile industry includes fluctuations in fuel prices	4.34	0.86	1.00	5.00
A_{16}	Availability of skilled labor at low cost is important growth driver of automobile industry	4.38	0.85	2.00	5.00
A_{17}	Robust Research and Development centers affects the growth of automobile industry	4.50	0.59	3.00	5.00
A_{18}	Greater fuel efficiency is important growth driver of automobile industry	4.36	0.76	2.00	5.00
A_{19}	New segment of customers are growth driver of automobile industry	4.23	0.67	2.00	5.00

Source: SPSS output from field survey data

Table 5.3 indicates that the mean value in all the cases is above 4, which indicate that the respondents reported their views above average which means agreed or strongly agree with the corresponding statements with a very low standard deviation in the range of 1 to 5.

Reliability Test – Cronbach's alpha

Cronbach's coefficient alpha is the best-known test of reliability used by numerous researches which will test the consistency of respondent answer to all the items in the measurement.

Table 5.4: Cronbach's Alpha Reliability Test

Reliability Statistics	
Cronbach's Alpha	N of Items
.775	12

Source: SPSS output from field survey data

Table No 5.4 indicates Cronbach's coefficient .775 which is an excellent measure of reliability test for the given statements under study.

Where the average is between the 2 groups we tested them by using t-test and in the case where the averages are more than 2 groups then we will use the F Test.

Table 5.5 shows the respondents' views according to their gender. A total number of 12 statements and views of all respondents reported on each statement are shown in the table below.

Table 5.5 shows that the perception level of the statements which is found to be more than 4 among the male and female category. Further, we used t-test to find out the significant difference in the average of 2 categories.

The table reports the average t statistics with significance values. The significance value (p value) is lower than 0.05 indicate that the difference in average is significant. The significance values ($p<0.05$) for A3, A12 and A13 clearly indicate that there is a significant difference in the perception level among male and female respondents.

Table 5.5: Wilcuxen *t*-test based on the views according to the gender of the respondents

Item	Statement	Gender	N	Mean	Standard Error	t-statistic	p-value	Remarks
A_1	The disclosure of return on equity affects the value of a Company.	Male	149	4.30	.087		.281	Insignificant
		Female	62	4.44	.097			
A_2	The takeover substantially affects the performance of a Company	Male	149	4.50	.054	.300	.764	Insignificant
		Female	62	4.47	.075			
A_3	The corporate takeover affects the interest of the stakeholders	Male	149	4.56	.047	2.070	.041	Significant at 5% level
		Female	62	4.35	.089			
A_4	Geographic coverage are the Company's primary reason for an acquisition	Male	149	4.32	.060		.542	Insignificant
		Female	62	4.39	.081			
A_5	The growth driver in takeover of a Company is operating synergy	Male	149	4.36	.055	.389	.698	Insignificant
		Female	62	4.32	.088			
A_6	The sharing of information about the future growth potentials plays an important role in the success of corporate takeover	Male	149	4.26	.066		.385	Insignificant
		Female	62	4.35	.080			
A_7	The primary reasons of failure of a corporate takeover include corporate cultural differences.	Male	149	4.28	.067	.866	.388	Insignificant
		Female	62	4.18	.093			
A_8	Proper communication among the stakeholders during the transition period is necessary to mitigate the	Male	149	4.47	.052	.546	.586	Insignificant
		Female	62	4.42	.071			

Item	Statement	Gender	N	Mean	Standard Error	t-statistic	p-value	Remarks
	failure of a corporate takeover.							
A 9	Measuring market trends for future growth potential of Company are important for success of corporate takeover.	Male	149	4.40	.055		.996	Insignificant
		Female	62	4.40	.084			
A 10	In case of takeover of a Company initially the productivity may drop temporarily as people take time to become familiar with new systems	Male	149	4.28	.075	.523	.602	Insignificant
		Female	62	4.21	.115			
A 11	India is an anchor of future growth in the auto industry	Male	149	4.39	.065	1.121	.263	Insignificant
		Female	62	4.26	.092			
A 12	The most popular form of vehicle in India are 2 wheelers / passenger cars	Male	149	4.46	.059	1.995	.047	**Significant at 5% level**
		Female	62	4.24	.097			
A 13	The factors that contribute to an increase in India's automobile demand is increasing buying power of customers	Male	149	4.55	.049	3.393	.001	**Significant at 5% level**
		Female	62	4.23	.090			
A 14	The automobile industry of tomorrow will reflect increased prosperity	Male	149	4.23	.078	.017	.986	Insignificant
		Female	62	4.23	.104			
A 15	The main challenges in the growth of Indian automobile industry includes	Male	149	4.27	.077		.076	Insignificant
		Female	62	4.50	.079			

Item	Statement	Gender	N	Mean	Standard Error	t-statistic	p-value	Remarks
	fluctuations in fuel prices							
A_{16}	Availability of skilled labor at low cost is important growth driver of automobile industry	Male	149	4.42	.071	.979	.329	Insignificant
		Female	62	4.29	.101			
A_{17}	Robust Research and Development centers affects the growth of automobile industry	Male	149	4.52	.050	.989	.324	Insignificant
		Female	62	4.44	.068			
A_{18}	Greater fuel efficiency is important growth driver of automobile industry	Male	149	4.39	.063	.863	.389	Insignificant
		Female	62	4.29	.093			
A_{19}	New segment of customers are growth driver of automobile industry	Male	149	4.19	.059		.267	Insignificant
		Female	62	4.31	.063			

Source: SPSS output from field survey data

Table 5.6 shows the respondent's views on effects, causes and growth drivers of M&A. A total number of 19 statements and views of all respondents reported on each statement are shown in the table below.

Table 5.6: *Respondents* views according to their age regarding affects of M&A

Item	Statement	Age Groups	N	Mean	Standard Error	F-statistic	p-value	Remarks
A1	The disclosure of return on equity affects the value of a Company.	<25 years	42	4.286	.109	2.114	.100	Insignificant
		25-35 years	66	4.288	.118			
		35-45 years	40	4.100	.185			
		>45 years	63	4.571	.125			
		Total	211	4.336	.067			
A2	The takeover substantially affects the performance of a Company	<25 years	42	4.429	.091	.154	.927	Insignificant
		25-35 years	66	4.500	.076			
		35-45 years	40	4.500	.113			
		>45 years	63	4.508	.081			
		Total	211	4.488	.044			
A3	The corporate takeover affects the interest of the stakeholders	<25 years	42	4.500	.104	.414	.743	Insignificant
		25-35 years	66	4.455	.069			
		35-45 years	40	4.475	.113			
		>45 years	63	4.571	.074			
		Total	211	4.502	.043			
A4	Geographic coverage are the Company's primary reason for an acquisition	<25 years	42	4.286	.104	1.157	.327	Insignificant
		25-35 years	66	4.303	.086			
		35-45 years	40	4.250	.133			
		>45 years	63	4.476	.078			
		Total	211	4.341	.048			

Item	Statement	Age Groups	N	Mean	Standard Error	F-statistic	p-value	Remarks
A 5	The growth driver in takeover of a Company is operating synergy	<25 years	42	4.429	.091	2.521	.059	Insignificant
		25-35 years	66	4.288	.088			
		35-45 years	40	4.150	.116			
		>45 years	63	4.492	.078			
		Total	211	4.351	.047			
A 6	The sharing of information about the future growth potentials plays an important role in the success of corporate takeover.	<25 years	42	4.119	.103	3.130	.027	Significant at 5% level
		25-35 years	66	4.303	.099			
		35-45 years	40	4.100	.128			
		>45 years	63	4.492	.087			
		Total	211	4.284	.052			
A 7	The primary reason of failure of a corporate takeover includes corporate cultural differences.	<25 years	42	4.310	.093	6.537	.000	Significant at 5% level
		25-35 years	66	3.924	.109			
		35-45 years	40	4.325	.131			
		>45 years	63	4.508	.087			
		Total	211	4.251	.055			
A 8	Proper communication among the stakeholders during the transition period is necessary to mitigate the failure of a corporate takeover.	<25 years	42	4.381	.096	1.433	.234	Insignificant
		25-35 years	66	4.364	.077			
		35-45 years	40	4.550	.094			
		>45 years	63	4.540	.074			
		Total	211	4.455	.042			

Item	Statement	Age Groups	N	Mean	Standard Error	F-statistic	p-value	Remarks
A 9	Measuring market trends for future growth potential of Company are important for success of corporate takeover.	<25 years	42	4.429	.114	2.409	.068	Insignificant
		25-35 years	66	4.303	.084			
		35-45 years	40	4.275	.107			
		>45 years	63	4.571	.071			
		Total	211	4.403	.046			
A 10	In case of takeover of a Company initially the productivity may drop temporarily as people take time to become familiar with new systems	<25 years	42	4.167	.107	1.211	.307	Insignificant
		25-35 years	66	4.136	.126			
		35-45 years	40	4.325	.145			
		>45 years	63	4.413	.115			
		Total	211	4.261	.063			
A 11	India is an anchor of future growth in the auto industry	<25 years	42	4.310	.125	2.105	.101	Insignificant
		25-35 years	66	4.182	.099			
		35-45 years	40	4.425	.138			
		>45 years	63	4.508	.078			
		Total	211	4.351	.053			
A 12	The most popular form of vehicle in India are 2 wheelers / passenger cars	<25 years	42	4.238	.136	3.434	.018	Significant at 5% level
		25-35 years	66	4.288	.091			
		35-45 years	40	4.375	.117			
		>45 years	63	4.635	.073			
		Total	211	4.398	.051			

Item	Statement	Age Groups	N	Mean	Standard Error	F-statistic	p-value	Remarks
A 13	The factors that contribute to an increase in India's automobile demand is increasing buying power of customers	<25 years	42	4.262	.103	4.843	.003	Significant at 5% level
		25-35 years	66	4.318	.089			
		35-45 years	40	4.575	.094			
		>45 years	63	4.651	.065			
		Total	211	4.455	.045			
A 14	The automobile industry of tomorrow will reflect increased prosperity	<25 years	42	4.167	.132	2.479	.062	Insignificant
		25-35 years	66	4.152	.111			
		35-45 years	40	4.025	.162			
		>45 years	63	4.476	.108			
		Total	211	4.227	.063			
A 15	The main challenges in the growth of Indian automobile industry includes fluctuations in fuel prices	<25 years	42	4.143	.130	2.561	.056	Insignificant
		25-35 years	66	4.303	.108			
		35-45 years	40	4.225	.150			
		>45 years	63	4.571	.098			
		Total	211	4.336	.060			
A 16	Availability of skilled labor at low cost is important growth driver of automobile industry	<25 years	42	4.476	.114	4.706	.003	Significant at 5% level
		25-35 years	66	4.227	.118			
		35-45 years	40	4.100	.142			
		>45 years	63	4.651	.085			
		Total	211	4.379	.059			

Item	Statement	Age Groups	N	Mean	Standard Error	F-statistic	p-value	Remarks
A17	Robust Research and Development centers affects the growth of automobile industry	<25 years	42	4.500	.085	6.445	.000	Significant at 5% level
		25-35 years	66	4.424	.075			
		35-45 years	40	4.250	.093			
		>45 years	63	4.730	.065			
		Total	211	4.498	.041			
A18	Greater fuel efficiency is important growth driver of automobile industry	<25 years	42	4.381	.108	.576	.631	Insignificant
		25-35 years	66	4.333	.097			
		35-45 years	40	4.250	.128			
		>45 years	63	4.444	.093			
		Total	211	4.360	.052			
A19	New segment of customers are growth driver of automobile industry	<25 years	42	4.214	.080	1.815	.146	Insignificant
		25-35 years	66	4.167	.077			
		35-45 years	40	4.100	.123			
		>45 years	63	4.381	.089			
		Total	211	4.227	.046			

Source: SPSS output from field survey data

The above results shows that the views of respondents with respect to the statement i.e. the sharing of information about the future growth potentials plays an important role in the success of corporate takeover, the primary reason of failure of a corporate takeover includes corporate cultural differences, the most popular form of vehicle in India are 2 wheelers and passenger cars, the factors that contribute to an increase in India's automobile demand is increasing buying power, Availability of skilled labor at low cost is an important growth driver of the automobile industry, Robust Researchand Development centers affects the growth of the automobile industry. The table indicates the significant variation at 5% level. It concludes that age wise respondent's view with respect to effects, cause, growth drivers of M&A are significant.

Table 5.7 shows that the perception level of the statements which is found to be more than 4 of the various professions of respondents. Further, we used the F Test to find out the significant difference in the average among the categories.

Table 5.7 shows the average F statistics with significance values. The significance value (p value) is lower than 0.5 indicates that the difference is significant. The significance values ($p<0.05$) for the disclosure of return on equity affects the value of a company, the takeover substantially affects the performance of a company, the corporate takeover affects the interest of the stakeholders, the primary reason of failure of a corporate takeover includes corporate cultural differences, proper communication among the stakeholders during the transition period is necessary to mitigate the failure of a corporate takeover, measuring market trends for future growth potential of company are important for success of corporate takeover, in case of takeover of a company initially the productivity may drop temporarily as people take time to become familiar with new systems, India is an anchor of future growth in the auto industry, the most popular form of vehicle in India are 2 wheelers and passenger cars, the factors that contribute to an increase in India's automobile demand is increasing buying power, the automobile industry of tomorrow will reflect increased prosperity, availability of skilled labor at low cost is an important growth driver of the automobile industry, a new segment of customers are growth driver of the automobile industry, clearly indicates that there is a significant difference in the views of the various professional groups of respondents.

Table 5.7 F Test on the profession wise distribution of respondents

Item	Statement	Age Groups	N	Mean	Standard Error	F-statistic	p-value	Remarks
A 1	The disclosure of return on equity affects the value of a Company.	Chartered Accountant	111	4.541	.082	3.567	.004	Significant at 5% level
		Company Secretary	65	4.046	.132			
		Cost Accountant	4	3.250	.750			
		Advocate	1	5.000				
		MBA	6	4.000	.516			
		Others	24	4.417	.169			
		Total	211	4.336	.067			
A 2	The takeover substantially affects the performance of a Company	Chartered Accountant	111	4.622	.057	2.640	.024	Significant at 5% level
		Company Secretary	65	4.354	.083			
		Cost Accountant	4	4.500	.289			
		Advocate	1	4.000				
		MBA	6	4.000	.000			
		Others	24	4.375	.132			
		Total	211	4.488	.044			
A 3	The corporate takeover affects the interest of the stakeholders	Chartered Accountant	111	4.631	.049	2.683	.023	Significant at 5% level
		Company Secretary	65	4.400	.087			
		Cost Accountant	4	4.500	.289			
		Advocate	1	4.000				
		MBA	6	4.000	.258			
		Others	24	4.333	.143			
		Total	211	4.502	.043			
A 4	Geographic coverage are the Company's primary reason for an acquisition	Chartered Accountant	111	4.387	.063	.467	.800	Insignificant
		Company Secretary	65	4.323	.093			
		Cost Accountant	4	4.250	.250			
		Advocate	1	4.000				

Item	Statement	Age Groups	N	Mean	Standard Error	F-statistic	p-value	Remarks
		MBA	6	4.000	.516			
		Others	24	4.292	.127			
		Total	211	4.341	.048			
A 5	The growth driver in takeover of a Company is operating synergy	Chartered Accountant	111	4.459	.062	1.957	.086	Insignificant
		Company Secretary	65	4.246	.088			
		Cost Accountant	4	4.250	.250			
		Advocate	1	3.000				
		MBA	6	4.333	.211			
		Others	24	4.208	.134			
		Total	211	4.351	.047			
A 6	The sharing of information about the future growth potentials plays an important role in the success of corporate takeover	Chartered Accountant	111	4.360	.071	1.613	.158	Insignificant
		Company Secretary	65	4.200	.099			
		Cost Accountant	4	3.750	.479			
		Advocate	1	3.000				
		MBA	6	4.000	.365			
		Others	24	4.375	.118			
		Total	211	4.284	.052			
A 7	The primary reason of failure of a corporate takeover includes corporate cultural differences.	Chartered Accountant	111	4.378	.077	2.471	.034	Significant at 5% level
		Company Secretary	65	4.031	.098			
		Cost Accountant	4	4.250	.250			
		Advocate	1	3.000				
		MBA	6	4.667	.211			
		Others	24	4.208	.147			
		Total	211	4.251	.055			
A 8	Proper communication among the stakeholders during the transition period is necessary to mitigate the failure of a corporate takeover.	Chartered Accountant	111	4.568	.054	3.087	.010	Significant at 5% level
		Company Secretary	65	4.338	.074			
		Cost Accountant	4	4.250	.250			
		Advocate	1	3.000				
		MBA	6	4.667	.333			
		Others	24	4.292	.141			

Item	Statement	Age Groups	N	Mean	Standard Error	F-statistic	p-value	Remarks
		Total	211	4.455	.042			
A 9	Measuring market trends for future growth potential of Company are important for success of corporate takeover.	Chartered Accountant	111	4.631	.056	8.162	.000	Significant at 5% level
		Company Secretary	65	4.154	.077			
		Cost Accountant	4	4.250	.250			
		Advocate	1	3.000				
		MBA	6	3.667	.422			
		Others	24	4.292	.127			
		Total	211	4.403	.046			
A 10	In case of takeover of a Company initially the productivity may drop temporarily as people take time to become familiar with new systems	Chartered Accountant	111	4.468	.078	3.024	.012	Significant at 5% level
		Company Secretary	65	3.985	.132			
		Cost Accountant	4	4.250	.250			
		Advocate	1	4.000				
		MBA	6	3.667	.422			
		Others	24	4.208	.134			
		Total	211	4.261	.063			
A 11	India is an anchor of future growth in the auto industry	Chartered Accountant	111	4.495	.069	3.417	.005	Significant at 5% level
		Company Secretary	65	4.246	.098			
		Cost Accountant	4	3.250	.250			
		Advocate	1	4.000				
		MBA	6	3.833	.477			
		Others	24	4.292	.141			
		Total	211	4.351	.053			
A 12	The most popular form of vehicle in India are 2 wheelers / passenger	Chartered Accountant	111	4.568	.062	3.429	.005	Significant at 5% level
		Company Secretary	65	4.277	.084			
		Cost	4	4.000	.000			

Item	Statement	Age Groups	N	Mean	Standard Error	F-statistic	p-value	Remarks
	cars	Accountant						
		Advocate	1	3.000				
		MBA	6	4.167	.543			
		Others	24	4.125	.193			
		Total	211	4.398	.051			
A 13	The factors that contribute to an increase in India's automobile demand is increasing buying power of customers	Chartered Accountant	111	4.577	.055	2.782	.019	Significant at 5% level
		Company Secretary	65	4.354	.089			
		Cost Accountant	4	4.250	.250			
		Advocate	1	3.000				
		MBA	6	4.500	.224			
		Others	24	4.250	.138			
		Total	211	4.455	.045			
A 14	The automobile industry of tomorrow will reflect increased prosperity	Chartered Accountant	111	4.468	.073	4.523	.001	Significant at 5% level
		Company Secretary	65	3.938	.126			
		Cost Accountant	4	3.500	.500			
		Advocate	1	3.000				
		MBA	6	3.667	.333			
		Others	24	4.208	.190			
		Total	211	4.227	.063			
A 15	The main challenges in the growth of Indian automobile industry includes fluctuations in fuel prices	Chartered Accountant	111	4.477	.074	2.099	.067	Insignificant
		Company Secretary	65	4.169	.111			
		Cost Accountant	4	3.500	.500			
		Advocate	1	4.000				
		MBA	6	4.000	.516			
		Others	24	4.375	.189			
		Total	211	4.336	.060			

Item	Statement	Age Groups	N	Mean	Standard Error	F-statistic	p-value	Remarks
A 16	Availability of skilled labor at low cost is important growth driver of automobile industry	Chartered Accountant	111	4.586	.072	3.271	.007	Significant at 5% level
		Company Secretary	65	4.215	.109			
		Cost Accountant	4	3.750	.479			
		Advocate	1	4.000				
		MBA	6	4.167	.477			
		Others	24	4.042	.185			
		Total	211	4.379	.059			
A 17	Robust Research and Development centers affects the growth of automobile industry	Chartered Accountant	111	4.568	.058	1.577	.168	Insignificant
		Company Secretary	65	4.446	.073			
		Cost Accountant	4	4.750	.250			
		Advocate	1	5.000				
		MBA	6	4.500	.224			
		Others	24	4.250	.090			
		Total	211	4.498	.041			
A 18	Greater fuel efficiency is important growth driver of automobile industry	Chartered Accountant	111	4.423	.068	.520	.761	Insignificant
		Company Secretary	65	4.277	.106			
		Cost Accountant	4	4.500	.500			
		Advocate	1	4.000				
		MBA	6	4.500	.224			
		Others	24	4.250	.138			
		Total	211	4.360	.052			
A 19	New segment of customers are growth driver of automobile industry	Chartered Accountant	111	4.378	.067	2.853	.016	Significant at 5% level
		Company Secretary	65	4.000	.073			
		Cost Accountant	4	4.250	.479			

Item	Statement	Age Groups	N	Mean	Standard Error	F-statistic	p-value	Remarks
		Advocate	1	4.000				
		MBA	6	4.167	.307			
		Others	24	4.167	.098			
		Total	211	4.227	.046			

Source: SPSS output from field survey data

Table 5.8 shows experience wise distribution of respondents views with respect to M&A. Further, we used the F Test to find out the significant difference in the average among the categories.

Table 5.8 F Test on the Experience wise distribution of Respondents views

Item	Statement	Age Groups	N	Mean	Standard Error	F-statistic	p-value	Remarks
A 1	The disclosure of return on equity affects the value of a Company.	< 1 year	33	3.970	.166	2.208	.069	Insignificant
		1-5 year	46	4.261	.126			
		6-10 years	28	4.571	.140			
		11-20 years	39	4.282	.176			
		>20 years	65	4.508	.130			
		Total	211	4.336	.067			
A 2	The takeover substantially affects the performance of a Company	< 1 year	33	4.152	.116	3.301	.012	**Significant at 5% level**
		1-5 year	46	4.609	.085			
		6-10 years	28	4.643	.092			
		11-20 years	39	4.487	.103			
		>20 years	65	4.508	.082			
		Total	211	4.488	.044			
A 3	The corporate takeover affects	< 1 year	33	4.242	.145	2.199	.070	Insignificant

	the interest of the stakeholders	1-5 year	46	4.565	.080			
		6-10 years	28	4.500	.096			
		11-20 years	39	4.462	.096			
		>20 years	65	4.615	.072			
		Total	211	4.502	.043			
A 4	Geographic coverage are the Company's primary reason for an Acquisition	< 1 year	33	4.152	.138	1.507	.201	Insignificant
		1-5 year	46	4.304	.098			
		6-10 years	28	4.429	.108			
		11-20 years	39	4.256	.141			
		>20 years	65	4.477	.073			
		Total	211	4.341	.048			
A 5	The growth driver in takeover of a Company is operating synergy	< 1 year	33	4.273	.125	3.492	.009	**Significant at 5% level**
		1-5 year	46	4.304	.093			
		6-10 years	28	4.607	.107			
		11-20 years	39	4.077	.118			

		>20 years	65	4.477	.079			
		Total	211	4.351	.047			
A 6	The sharing of information about the future growth potentials plays an important role in the success of corporate takeover	< 1 year	33	3.909	.133	5.215	.001	**Significant at 5% level**
		1-5 year	46	4.500	.087			
		6-10 years	28	4.393	.149			
		11-20 years	39	4.026	.135			
		>20 years	65	4.431	.088			
		Total	211	4.284	.052			
A 7	The primary reasons of failure of a corporate takeover includes corporate cultural differences.	< 1 year	33	4.152	.138	2.884	.024	**Significant at 5% level**
		1-5 year	46	3.957	.124			
		6-10 years	28	4.286	.144			
		11-20 years	39	4.333	.124			
		>20 years	65	4.446	.093			
		Total	211	4.251	.055			
A 8	Proper communication	< 1 year	33	4.273	.117	1.144	.337	Insignificant

	among the stakeholders during the transition period is necessary to mitigate the failure of a corporate takeover.	1-5 year	46	4.413	.091			
		6-10 years	28	4.536	.096			
		11-20 years	39	4.487	.103			
		>20 years	65	4.523	.073			
		Total	211	4.455	.042			
A 9	Measuring market trends for future growth potential of Company are important for success of corporate takeover.	< 1 year	33	4.242	.138	1.754	.139	Insignificant
		1-5 year	46	4.370	.100			
		6-10 years	28	4.393	.130			
		11-20 years	39	4.308	.111			
		>20 years	65	4.569	.066			
		Total	211	4.403	.046			
A 10	In case of takeover of a Company initially the productivity may drop temporarily as people take time to become familiar with new systems	< 1 year	33	3.970	.141	2.095	.083	Insignificant
		1-5 year	46	4.109	.143			
		6-10 years	28	4.536	.131			
		11-20 years	39	4.385	.150			

		>20 years	65	4.323	.118			
		Total	211	4.261	.063			
A 11	India is an anchor of future growth in the auto industry	< 1 year	33	4.182	.154	3.002	.020	Significant at 5% level
		1-5 year	46	4.109	.117			
		6-10 years	28	4.571	.108			
		11-20 years	39	4.333	.148			
		>20 years	65	4.523	.076			
		Total	211	4.351	.053			
A 12	The most popular form of vehicle in India are 2 wheelers / passenger cars	< 1 year	33	4.152	.158	2.905	.023	Significant at 5% level
		1-5 year	46	4.217	.107			
		6-10 years	28	4.500	.141			
		11-20 years	39	4.436	.109			
		>20 years	65	4.585	.079			
		Total	211	4.398	.051			
A 13	The factors that contribute to an	< 1 year	33	4.121	.136	8.270	.000	Significant at 5%

	increase in India's automobile demand is increasing buying power	1-5 year	46	4.174	.095			level
		6-10 years	28	4.750	.083			
		11-20 years	39	4.564	.103			
		>20 years	65	4.631	.064			
		Total	211	4.455	.045			
A 14	The automobile industry of tomorrow will reflect increased prosperity	<1 year	33	4.091	.153	1.179	.321	Insignificant
		1-5 year	46	4.087	.128			
		6-10 years	28	4.321	.171			
		11-20 years	39	4.154	.158			
		>20 years	65	4.400	.114			
		Total	211	4.227	.063			
A 15	The main challenges in the growth of Indian Automobile Industry includes fluctuations in fuel prices	<1 year	33	3.970	.166	3.491	.009	**Significant at 5% level**
		1-5 year	46	4.217	.124			
		6-10 years	28	4.536	.141			
		11-20 years	39	4.256	.159			

		>20 years	65	4.569	.090			
		Total	211	4.336	.060			
A 16	Availability of skilled labor at low cost is important growth driver of automobile Industry	< 1 year	33	4.394	.130	2.738	.030	**Significant at 5% level**
		1-5 year	46	4.196	.151			
		6-10 years	28	4.393	.149			
		11-20 years	39	4.154	.145			
		>20 years	65	4.631	.087			
		Total	211	4.379	.059			
A 17	Robust Research and Development centers affects the growth of automobile Industry	< 1 year	33	4.515	.088	2.923	.022	**Significant at 5% level**
		1-5 year	46	4.370	.095			
		6-10 years	28	4.500	.121			
		11-20 years	39	4.333	.076			
		>20 years	65	4.677	.073			
		Total	211	4.498	.041			
A 18	Greater fuel efficiency is	< 1 year	33	4.303	.111	.692	.598	Insignificant

		important growth driver of automobile Industry	1-5 year	46	4.391	.118			
			6-10 years	28	4.393	.149			
			11-20 years	39	4.205	.133			
			>20 years	65	4.446	.090			
			Total	211	4.360	.052			
A 19	New segment of customers are growth driver of automobile industry		< 1 year	33	4.152	.088	.588	.672	Insignificant
			1-5 year	46	4.152	.076			
			6-10 years	28	4.214	.140			
			11-20 years	39	4.231	.119			
			>20 years	65	4.323	.093			
			Total	211	4.227	.046			

Source: SPSS output from field survey data

The above Table no. 5.8 indicates F statistics with significance values. The significance value (p value) is lower than 0.5 indicate that the difference in average is significant. The significance values ($p<0.05$) for the takeover substantially affects the performance of a company, the growth driver in takeover of a company is operating synergy, the sharing of information about the future growth potentials plays an important role in the success of corporate takeover, the primary reasons for failure of a corporate takeover includes corporate cultural differences, India is an anchor of future growth in the auto industry, the most popular form of vehicle in India are 2 wheelers and passenger cars, the factors that contribute to an increase in India's automobile demand is increasing buying power, the main challenges in the growth of the Indian automobile industry includes fluctuations in fuel prices, availability of skilled labor at low cost is an important growth driver of the automobile industry, and Robust Research and development centers affect the growth of the automobile industry, clearly indicate that there is a significant difference in the views of different age groups of respondents.

KMO & Bartlett's test

KMO test of the sampling adequacy and Bartlett's test of Sphericity were performed to confirm the suitability of the data for factor analysis.

Table 5.9: KMO & Bartlett's test

Kaiser-Meyer-Olkin Measure of Sampling Adequacy.		.863
Bartlett's Test of Sphericity	Approx. Chi-Square	1047.175
	Df	171
	Sig.	.000

Source: SPSS output from field survey data

The table no. 5.9 shows the suitability of sampling adequacy i.e. 0.863, which is suitable for factor analysis.

Factor Analysis

Factor analysis performed on nineteen statements and results shown in table no 5.10. It shows that the statement numbers A1, A14, A15, and A19 have been loaded on component 1. Whereas statement numbers A5, A6, A12, and A13 have been loaded on component 2. Further statement numbers A16, A17 and A18 have been loaded on component 3. The statement numbers A4, A6, A8, A9 and A11 have been loaded on component 4 and lastly, the statement numbersA2, and A3 loaded on component 5.

Table 5.10: Rotated Component Matrix[a]

	Component				
	1	2	3	4	5
A_1	0.670	0.238	0.024	0.013	0.160
A_2	0.109	-0.007	0.109	0.301	0.754
A_3	0.355	0.315	0.048	-0.281	0.541
A_4	0.288	0.123	0.341	0.523	-0.126
A_5	-0.006	0.505	0.083	0.184	0.405
A_6	0.437	0.013	0.153	0.515	0.298
A_7	0.384	0.575	0.174	-0.008	0.008
A_8	0.029	**0.490**	0.008	0.584	0.102
A_9	0.258	**0.140**	0.036	0.615	0.255
A_{10}	0.512	0.267	-0.028	0.286	0.064
A_{11}	0.231	0.362	0.108	0.522	0.187
A_{12}	0.241	0.620	0.080	0.246	-0.129
A_{13}	0.131	0.754	0.093	0.113	0.143
A_{14}	0.681	0.135	-0.039	0.240	0.067
A_{15}	0.630	0.070	-0.015	0.447	-0.110
A_{16}	0.170	0.114	0.814	0.120	0.013
A_{17}	0.043	0.205	0.771	-0.140	0.037
A_{18}	-0.072	-0.033	0.768	0.260	0.169
A_{19}	0.698	0.079	0.242	0.045	0.116

Source: SPSS output from field survey data

Extraction Method: Principal Component Analysis.

Rotation Method: Varimax with Kaiser Normalization.

a. Rotation converged in 25 iterations

Total variance explained in Table no. 5.11 shows that 56.17 percent of total variance explained by 5 components.

Table 5.11: Total Variance Explained

Component	Initial Eigen values			Rotation Sums of Squared Loadings		
	Total	Percentage of Variance	Cumulative Percentage	Total	Percentage of Variance	Cumulative Percentage
1	5.403	28.435	28.435	2.829	14.891	14.891
2	1.819	9.576	38.011	2.247	11.826	26.717
3	1.243	6.541	44.552	2.127	11.193	37.91
4	1.161	6.109	50.661	2.103	11.07	48.98
5	1.048	5.516	56.177	1.367	7.197	56.177
6	0.907	4.774	60.951			
7	0.868	4.567	65.518			
8	0.788	4.146	69.665			
9	0.738	3.882	73.547			
10	0.656	3.453	77			
11	0.601	3.165	80.165			
12	0.576	3.03	83.195			
13	0.549	2.889	86.084			
14	0.535	2.817	88.9			
15	0.514	2.705	91.605			
16	0.448	2.356	93.961			
17	0.431	2.267	96.228			
18	0.39	2.051	98.279			
19	.327	1.721	100			

Source: SPSS output from field survey data

Extraction Method: Principal Component Analysis.

The following Rotated matrix can also be used

Table 5.12: Rotated Component Matrix[a]

	Component				
	1	2	3	4	5
A_1	0.67	0.238	0.024	0.013	0.16
A_{10}	0.512	0.267	-0.028	0.286	0.064
A_{14}	0.681	0.135	-0.039	0.24	0.067
A_{15}	0.63	0.07	-0.015	0.447	-0.11
A_{19}	0.698	0.079	0.242	0.045	0.116
A_5	-0.006	0.505	0.083	0.184	0.405
A_7	0.384	0.575	0.174	-0.008	0.008
A_{12}	0.241	0.62	0.08	0.246	-0.129
A_{13}	0.131	0.754	0.093	0.113	0.143
A_{16}	0.170	0.114	0.814	0.120	0.013
A_{17}	0.043	0.205	0.771	-0.140	0.037
A_{18}	-0.072	-0.033	0.768	0.260	0.169
A_4	0.288	0.123	0.341	0.523	-0.126
A_6	0.437	0.013	0.153	0.515	0.298
A_8	0.029	**0.49**	0.008	0.584	0.102
A_9	0.258	**0.14**	0.036	0.615	0.255
A_{11}	0.231	0.362	0.108	0.522	0.187
A_2	0.109	-0.007	0.109	0.301	0.754
A_3	0.355	0.315	0.048	-0.281	0.541

Source: SPSS output from field survey data

Extraction Method: Principal Component Analysis.

Rotation Method: Varimax with Kaiser Normalization.

a. Rotation converged in 25 iterations.

On the basis of above table No. 5.12 it can be stated that the above statements can be grouped into the following five components and are considered as major factors affecting the takeover of companies under automobile industry:

Table 5.13: Results of Factor analysis

S. No.	Item	Statement Description	Nomenclature	Factor identified
1	A 1	The disclosure of return on equity affects the value of a Company.	For a successful takeover, necessary factors are a new segment of customers, the greater fuel efficiency of vehicles and better return on equity.	Value of Company
	A 10	In case of takeover of a Company initially the productivity may drop temporarily as people take time to become familiar with new systems		
	A 14	The automobile industry of tomorrow will reflect increased prosperity		
	A 15	The main challenges in the growth of Indian automobile industry includes fluctuations in fuel prices		
	A 19	New segment of customers are growth driver of the automobile industry		
2	A 5	The growth driver in the takeover of a Company is operating synergy	Growth drivers of the takeovers are operating synergy, increased buying power of the people and popularity of two-wheeler and passenger cars, subject to the management of corporate cultural differences.	Managing growth during the transition period
	A 7	The primary reasons of failure of a corporate takeover include corporate cultural differences.		
	A 12	The most popular form of vehicle in India are two wheelers and passenger cars		
	A 13	The factors that contribute to an increase in India's automobile demand is increasing buying		

S. No.	Item	Statement Description	Nomenclature	Factor identified
		power of customers		
3	A 16	Availability of the skilled labor at low cost is an important growth driver of automobile industry	The growth drivers of automobile industry include the availability of skilled labor at low cost, robust Research and Development centers and greater fuel efficiency.	Growth Drivers
	A 17	Robust Research and Development centers affect the growth of the automobile industry		
	A 18	Greater fuel efficiency is important growth driver of the automobile industry		
4	A 4	Geographic coverage is the Company's primary reason for an acquisition	Proper communication among the stakeholders during the corporate transition period, the intent of geographical expansion and sharing of information about the future growth potentials are important reasons behind successful corporate takeovers.	Future plans and Geographical expansion
	A 6	The sharing of information about the future growth potentials plays an important role in the success of the corporate takeover		
	A 8	Proper communication among the stakeholders during the transition period is necessary to mitigate the failure of a corporate takeover.		
	A 9	Measuring market trends for the future growth potential of Company are important for the success of corporate takeover.		
	A 11	India is an anchor of future growth in the auto industry		
5	A 2	The takeover substantially affects the performance of a Company	The primary reasons for corporate takeovers are to improve the performance of the company and to improve the interest of its stakeholders.	Stakeholder's interest
	A 3	The corporate takeover affects the interest of the stakeholders		

Based on the above tests and analysis the following key factors have been identified:

i. Value of Company
ii. Managing growth during the transition period
iii. Growth Drivers
iv. Future plans and Geographical expansion
v. Stakeholder's interest

5.2 Analysis of Secondary Data

In the following sections, the pre and post acquisition scenario of the following four companies has been analyzed. The pre and post identities of these companies have remained unchanged.

List of selected Companies

S. No.	Name of the Company
1	TATA Motors Ltd.
2	Volkswagen
3	Mahindra & Mahindra Ltd.
4	Daimler Benz

The broad objective of the secondary data based study is to examine the impact of the corporate takeover on the growth and financial performance of selected companies under the automotive industry.

Annual reports and accounts of the selected companies are the sources of the data used in this study. Descriptive and t-test statistics computed using SPSS used to analyze the collected data.

TATA MOTORS Ltd.

Table 5.14: Operational and Financial performance of TATA Motors Ltd. during the pre and post takeover period

(Rs in crores)

Particulars	2003	2004	2005	2006	2007	2008	2009	2010	2011	2012	2013
		Pre Merger				Year of merger			Post Merger		
Net Sales	9,612	13,925	19,440	23,588	32,067	35,413	71,608	91,700	122,128	165,654	188,818
Other Income	19	56	183	351	430	653	(800)	3,945	429	662	812
Total Revenue	**9,631**	**13,981**	**19,623**	**23,939**	**32,497**	**36,066**	**70,808**	**95,645**	**122,557**	**166,316**	**189,629**
Less: Direct Exp.	(5,915)	(9,396)	(14,226)	(17,146)	(23,554)	(26,173)	(50,335)	(64,733)	(80,006)	(110,857)	(122,343)
Gross Profit	**3,716**	**4,585**	**5,397**	**6,793**	**8,943**	**9,893**	**20,472**	**30,912**	**42,551**	**55,459**	**67,286**
% of Gross Profit to Sales	39%	33%	28%	29%	28%	28%	29%	34%	35%	33%	36%
Less: Indirect and Extraordinary Exp.	(2,445)	(2,526)	(2,852)	(3,575)	(4,763)	(5,282)	(18,164)	(21,376)	(25,074)	(33,318)	(42,530)
Earnings before Depreciation, Interest & Taxes (EBDIT)	1,271	2,059	2,546	3,217	4,180	4,611	2,308	9,536	17,478	22,141	24,756
% of EBDIT to Sales	13%	15%	13%	13%	13%	13%	3%	10%	14%	13%	13%
Less: Interest	(325)	(194)	(170)	(246)	(406)	(743)	(1,931)	(2,126)	(2,385)	(2,982)	(3,553)
Less: Depreciation	(402)	(426)	(531)	(623)	(688)	(782)	(2,507)	(3,887)	(4,656)	(5,625)	(7,569)
Profit before Taxes (PBT)	**544**	**1,439**	**1,845**	**2,348**	**3,086**	**3,086**	**(2,129)**	**3,523**	**10,437**	**13,534**	**13,633**
% of PBT to Sales	6%	10%	9%	10%	10%	9%	(3%)	4%	9%	8%	7%
Less: Tax	(226)	(531)	(491)	(640)	(883)	(852)	(336)	(1,006)	(1,216)	40	(3,771)
Profit after Taxes (PAT)	**317**	**909**	**1,354**	**1,708**	**2,203**	**2,235**	**(2,465)**	**2,517**	**9,221**	**13,574**	**9,862**
% of PAT to Sales	3%	7%	7%	7%	7%	6%	(3%)	3%	8%	8%	5%
Earnings Per Share (EPS)	9	26	38	45	56	56	(49)	45	31	43	31

Sources: Data compiled from the official website of Tata Motors Ltd. and Money control

Table no. 5.14 shows the pre and post takeover operational and financial performance of TATA Motors Ltd. during the years from 2003 to 2013. It indicates the fluctuations in PBT (profit before tax), PAT (profit after tax), and EPS (earnings per share) during pre and post takeover scenario.

Table 5.15: Descriptive test statistics based on profitability ratios - pre and post takeover scenario

	Gross Profit Ratio		Net Profit Ratio		Earnings per Share (EPS)	
	Pre	Post	Pre	Post	Pre	Post
MAX	3.87	3.56	10.00	7.00	60.00	34.00
MIN	2.78	2.86	3.30	(3.44)	9.00	(49.00)
Mean	3.68	3.33	8.49	3.51	51.80	13.60
Std. Dev.	4.39	2.75	2.33	4.19	19.48	35.32

Source: t-test output from SPSS

Table 5.15 shows that gross profit ratio mean values are 3.68 for pre takeover and 3.33 for post takeover, with a minimum value of 2.78 for pre takeover and 2.86 for the post takeover, and the maximum value of 3.87 for pre takeover and 3.56 for the post takeover. The standard deviation of pre is 4.39 and post takeover is 2.75. The change of standard deviation shows that the company varies in their gross profit ratio before the takeover and after the takeover.

Net profit ratio mean values are 8.49 and 3.51 for pre acquisition and post acquisition, with a minimum value of 3.30 for pre takeover and (3.44) for the post takeover, and the maximum value of 10.00 for pre acquisition and 7.00 for the post acquisition. The standard deviation of pre is 2.33 and for the post takeover is 4.19. The change of standard deviation shows that the company varies in their net profit ratio before the takeover and after the takeover.

Earnings per share mean values are 51.80 for pre acquisition and 13.60 for the post acquisition, with a minimum value of 9.00 for pre acquisition and (49.00) for the post acquisition, and the maximum of 60.00 for pre acquisition and 34.00 for the post acquisition. The standard deviation of pre is 19.48 and post merger is 35.32. The growth of standard deviation shows that the company varies in their earnings per share before the takeover and after the takeover.

Table 5.16: t-test on profitability ratios

	Gross Profit Ratio	Net Profit Ratio	Earnings Per Share
Mean Correlation	(0.93)	0.80	0.86
df (Degree of freedom)	4.00	4.00	4.00
SIGN. (2 Tail)	0.02	0.13	0.18
t Test	0.04	0.03	0.04

Source: t-test output from SPSS

Table 5.16 shows that there is a significant difference in gross profit ratio, net profit ratio, and earnings per share during pre and post takeover scenario of TATA Motors Ltd.

The study concludes that TATA Motors Ltd's operational and financial performance has been significantly affected due to its takeover activity.

Table 5.17: Financial position of TATA Motors Ltd. during pre and post takeovers scenario

	2004	2005	2006	2007	2008	2009	2010	2011	2012	2013
	Pre Merger				Year of merger	Post Merger				
Current Ratio	0.89	1.13	1.30	1.27	1.02	0.56	0.71	0.76	0.88	0.86
Quick Ratio	0.49	0.84	1.01	1.03	0.84	0.38	0.52	0.51	0.63	0.65
Total Debt to Equity Ratio	0.46	0.59	0.43	0.48	0.69	1.72	1.97	1.26	1.09	1.15
Return on Capital Employed (%)	15.42	17.66	17.48	16.78	12.78	(14.08)	9.48	20.20	18.74	11.81

Sources: *Data compiled from the official website of Tata Motors Ltd. and Money control*

Table no. 5.17 shows TATA Motors financial position statement for the period 2004 to 2013. The current ratio of TATA Motors Ltd. in the year 2004 is 0.89 which increased to 1.27 in the year 2007 during pre takeover scenario. During post takeover scenario current ratio of TATA Motors Ltd. in the year 2009 reported at 0.56 which further increased to 0.86 in the year 2013. Similarly, its return on capital employed percentage in the year 2004 is 15.42 which increased to 16.78 in the year 2007 during pre takeover scenario. During post acquisition scenario return on capital employed percentage of TATA Motors Ltd. in the year 2009 reported at (14.08) which further increased to 11.81 in the year 2013.

Table 5.18: Descriptive test statistics based on liquidity ratios- pre and post takeovers of TATA Motors Ltd.

	Current Ratio		Quick Ratio	
	Pre	Post	Pre	Post
MAX	1.30	0.88	1.03	0.63
MIN	1.02	0.56	0.84	0.38
Mean	1.18	0.73	0.93	0.51
Std. Dev.	0.131	0.135	0.101	0.102

Source: t-test output from SPSS

Table 5.18 shows that current ratio mean values are 1.18 for pre takeover and 0.73 for the post takeover, with a minimum value of 1.02 for pre takeover and 0.56 for the post takeover, and the maximum value of 1.30 for pre takeover and 0.88 for the post takeover. The standard deviation of pre is 0.131 and post takeover is 0.135. The change of the standard deviation shows that the company varies in their current ratio before the takeover and after the takeover.

Quick ratio mean values are 0.93 for pre takeover and 0.51 for the post takeover, with a minimum value of 0.84 for pre takeover and 0.38 for the post takeover, and the maximum value of 1.03 for pre takeover and 0.63 for the post takeover. The standard deviation of pre is 0.101 and post takeover is 0.102. The change of the standard deviation shows that the company varies in their quick ratio before the takeover and after the takeover.

Table 5.19: t-test on Liquidity ratios

	Current Ratio	Quick Ratio
Mean Correlation	(0.29)	0.04
df (Degree of freedom)	3	3
SIGN. (2 Tail)	0.003	0.0029
t-test	0.024	0.009

Source: t-test output from SPSS

Table 5.19 shows that there is a significant difference with respect to the current ratio and quick ratio of Tata motors during its pre and post takeover scenario.

The study concludes that TATA Motors liquidity position is significantly affected due to its takeover activity.

Table 5.20: Descriptive test statistics based on Leverage ratios- pre and post takeover of TATA Motors Ltd.

	Total Debt to Equity Ratio		Return on Capital Employed (%)	
	Pre	Post	Pre	Post
MAX	0.69	1.97	17.66	20.2
MIN	0.43	1.09	12.78	(14.08)
Mean	0.55	1.51	16.18	8.59
Std.Dev.	0.113	0.404	2.30	15.84

Source: t-test output from SPSS

Table 5.20 shows that total debt to equity ratio mean values are 0.55 for pre takeover and 1.51 for the post takeover, with a minimum value of 0.43 for pre takeover and 1.09 for post takeover, and the maximum value of 0.69 for pre takeover and 1.97 for the post takeover. The standard deviation of pre is 0.113 and post is 0.404. The change of the standard deviation shows that the company varies in their total debt to equity ratio before takeover and after takeover.

Return on capital employed percent mean values are 16.18 for pre takeover and 8.59 for the post takeover, with a minimum value of 12.78 for pre takeover and (14.08) for the post takeover, and the maximum value of 17.66 for pre takeover and 20.20 for the post takeover. The standard deviation of pre is 2.30 and for the post takeover is 15.84. The change of the standard deviation shows that the company varies in their return on capital employed before takeover and after takeover.

Table 5.21: t-test on Leverage ratios

	Total Debt to Equity Ratio	Return on Capital Employed (%)
Mean Correlation	(0.64)	(0.55)
Df (Degree of freedom)	3	3
SIGN. (2 Tail)	0.0028	0.31
t-test	0.0289	0.32

Source: t-test output from SPSS

Table 5.21 shows a significant difference with respect to total debt to equity ratio of Tata motors. However, in the case of return on capital employed of Tata Motors Ltd. there is no significant difference during the pre and post takeover scenario.

Finding on the operational and financial performance of TATA Motors Ltd. during its pre and post acquisition scenario.

The above results show that the financial and operational performance of TATA Motors Ltd. with respect to its profitability and liquidity position, there is a significant difference between pre and post acquisition scenario. Further, the result of leverage ratios of Tata Motors Ltd. also found significant during pre and post takeover scenario except its return on capital employed

Therefore, based upon the above test results, it is concluded that corporate takeover has an impact on the growth and financial performance of TATA Motors Ltd.

VOLKSWAGEN

Table 5.22: Operational and Financial performance of VOLKSWAGEN during the pre and post takeovers period

(€ million)

Particulars	2007	2008	2009	2010	2011	2012	2013	2014	2015	2016	2017
		Pre Merger				Year of Merger			Post Merger		
Sales Revenue	108,897	113,808	105,187	126,875	159,337	192,676	197,007	202,458	213,292	217,267	230,682
Other Income	8,033	10,860	9,577	11,645	19,429	27,031	13,079	15,053	18,065	16,193	16,911
Total Revenue	116,930	124,668	114,764	138,520	178,766	219,707	210,086	217,511	231,357	233,460	247,593
Less: Direct Exp.	(97,013)	(102,951)	(97,960)	(111,881)	(138,827)	(166,588)	(168,750)	(172,926)	(199,553)	(193,177)	(200,399)
Gross Profit	19,917	21,717	16,804	26,639	39,939	53,119	41,336	44,585	31,804	40,283	47,194
Less: Indirect and Extraordinary Exp.	(11,727)	(13,294)	(13,276)	(15,500)	(18,966)	(25,073)	(26,543)	(27,133)	(30,712)	(30,036)	(30,964)
Profit before Interest & Taxes (PBIT)	8,190	8,423	3,528	11,139	20,973	28,046	14,793	17,452	1,092	10,247	16,230
Less: Interest	(1,647)	(1,815)	2	(2,144)	(2,047)	(2,552)	(2,366)	(2,658)	(2,393)	(2,955)	(2,317)
Profit before Taxes (PBT)	6,543	6,608	3,530	8,995	18,926	25,494	12,427	14,794	(1,301)	7,292	13,913
% of PBT to Sales	0	0	0	0	0	0	0	0	(0)	0	0
Less: Tax	(2,421)	(1,920)	(349)	(1,767)	(3,126)	(3,608)	(3,283)	(3,726)	(59)	(1,912)	(2,275)
Profit after Taxes (PAT)	4,122	4,688	3,181	7,228	15,800	21,886	9,144	11,068	(1,360)	5,380	11,638
% of PAT o Sales	0	0	0	0	0	0	0	0	(0)	0	0
Earnings per Share	10	12	2	15	33	46	19	22	(3)	10	23

Sources: Data compiled from the official website of Volkswagen and Money control

Table no. 5.22 shows pre and post takeover operational and financial performance of VOLKSWAGON during the years from 2007 to 2017. It indicates the fluctuations in PBT (profit before tax), PAT (profit after tax), and EPS (earnings per share) during pre and post takeover scenario.

Table 5.23: Descriptive test statistics based on profitability ratios - pre and post takeover scenario

	Gross Profit Ratio		Net Profit Ratio		Earnings Per Share	
	Pre	Post	Pre	Post	Pre	Post
MAX	0.25	0.22	0.10	0.05	33.10	22.63
MIN	0.16	0.15	0.03	(0.01)	2.37	(3.20)
Mean	0.20	0.19	0.05	0.03	14.60	14.02
Std.Dev.	0.034	0.028	0.028	0.025	11.37	10.81

Source: t-test Output from SPSS

Table 5.23 shows that gross profit ratio mean values are 0.20 for pre takeover and 0.19 for the post takeover, with a minimum value of 0.16 for pre takeover and 0.15 for the post takeover, and the maximum value of 0.25 for pre takeover and 0.22 for the post takeover. The standard deviation of pre is 0.034 and post takeover is 0.028. The change of standard deviation shows that company varies in their gross profit ratio before the takeover and after the takeover.

Net profit ratio mean values are 0.05 and 0.03 for pre takeover and post takeover, with a minimum value of 0.03 for pre takeover and (0.01) for the post takeover, and the maximum of 0.10 for pre takeover and 0.05 for the post takeover. The standard deviation of pre is 0.028 and for the post takeover is 0.025. The change of standard deviation shows that company varies in their net profit ratio before the takeover and after the takeover.

Earnings per share mean values are 14.60 for pre takeover and 14.02 for the post takeover, with a minimum value of 2.37 for pre takeover and (3.20) for the post takeover and the maximum of 33.10 for pre takeover and 22.63 for the post takeover.

The standard deviation of pre is 11.37 and for the post takeover is 10.81. The changes of standard deviation show that company varies in their earnings per share before the takeover and after the takeover.

Table 5.24: t-test on Profitability ratios

	Gross Profit Ratio	Net Profit Ratio	Earnings Per Share
Mean Correlation	0.46	0.43	0.67
df (Degree of freedom)	4.00	4.00	4.00
SIGN. (2 Tail)	0.81	0.29	0.94
t-test	0.75	0.20	0.89

Source: t-test Output from SPSS

Table 5.24 shows that there is no significant difference in gross profit ratio, net profit ratio and earnings per share during pre and post takeover scenario of VOLKSWAGEN. The study concludes that VOLKSWAGEN's operational and financial performance has not been significantly affected due to its takeover activity.

Table 5.25: Financial position of VOLKSWAGEN during pre and post takeover scenario

	2007	2008	2009	2010	2011	2012	2013	2014	2015	2016	2017
	Pre Merger					Year of Merger	Post Merger				
Current Ratio	1.22	1.18	1.12	1.12	1.04	1.07	1.03	1.00	0.98	0.88	1.00
Quick Ratio	0.97	0.96	0.92	0.89	0.77	0.80	0.79	0.76	0.77	0.70	0.80
Total Debt to Equity Ratio	1.80	1.76	1.88	1.51	1.41	1.49	1.28	1.44	1.64	1.50	1.40
Return on Capital Employed (%)	6.89	6.14	1.72	5.83	7.39	5.64	5.67	5.76	5.44	3.06	5.28

Source: Data compiled from the official website of Volkswagen and Money control

Table no. 5.25 shows VOLKSWAGEN's financial position statement for the period 2007 to 2017. It shows that the current ratio of VOLKSWAGEN in the year 2007 is 1.22 which decreased to 1.04 in the year 2011 during pre takeover scenario. During post takeover scenario current ratio of VOLKSWAGEN in the year 2013 reported

1.03 which further decreased to 1.00 in the year 2017; similarly, its return on capital employed percentage in the year 2007 is 6.89, which increase to 7.39 in the year 2011 during pre takeover scenario. During post takeover scenario return on capital employed of VOLKSWAGEN in the year 2013 reported at 5.67 which further decreased to 5.28 in the year 2017.

Table 5.26: Descriptive test statistics based on liquidity ratios- pre and post takeover of VOLKSWAGEN

	Current ratio		Quick Ratio	
	Pre	Post	Pre	Post
MAX	1.22	1.03	0.97	0.80
MIN	1.04	0.88	0.77	0.70
Mean	1.14	0.98	0.90	0.76
Std. Dev.	0.067	0.059	0.079	0.039

Source: t-test Output from SPSS

Table 5.26 shows that current ratio mean values are 1.14 for pre takeover and 0.98 for the post takeover, with a minimum value of 1.04 for pre takeover and 0.88 for the post takeover and the maximum value of 1.22 for pre takeover and 1.03 for the post takeover. The standard deviation of pre is 0.067 and for the post takeover is 0.059. The changes of standard deviation show that company varies in their current ratio before the takeover and after the takeover.

Quick ratio mean values are 0.90 for pre takeover and 0.76 for the post takeover, with a minimum value of 0.77 for pre takeover and 0.70 for post takeover and the maximum value of 0.97 for pre takeover and 0.80 for the post takeover. The standard deviation of pre is 0.079 and for the post takeover is 0.039. The change of standard deviation shows that company varies in their quick ratio before the takeover and after the takeover.

Table 5.27: t-test on Liquidity ratios

	Current Ratio	Quick Ratio
Mean Correlation	0.34	-
df (Degree of freedom)	4.00	4.00
SIGN. (2 Tail)	0.004	0.01
t-test	0.01	0.03

Source: t-test Output from SPSS

Table 5.27 shows that there is a significant difference with respect to current ratio and quick ratio of VOLKSWAGEN during its pre and post takeover scenario. The study concludes that VOLKSWAGEN's liquidity position has been significantly affected due to its takeover activity.

Table 5.28: Descriptive test statistics based on leverage ratios- pre and post takeover of VOLKSWAGEN

	Total Debt to Equity ratio		Return on capital	
	Pre	Post	Pre	Post
MAX	1.88	1.64	0.07	0.06
MIN	1.41	1.28	0.02	0.03
Mean	1.67	1.45	0.06	0.05
Std. Dev.	0.199	0.132	0.022	0.011

Source: t-test Output from SPSS

Table 5.28 shows that total debt to equity ratio mean values are 1.67 for pre takeover and 1.45 for post takeover, with a minimum value of 1.41 for pre takeover and 1.28 for the post takeover and the maximum value of 1.88 for pre takeover and 1.64 for the post takeover. The standard deviation of pre is 0.199 and for the post is 0.132. The change of standard deviation shows that company varies in their total debt to equity ratio before the takeover and after the takeover.

Return on capital employed percent mean values are 0.06 for pre takeover and 0.05 for post the takeover, with a minimum value of 0.02 for pre takeover and 0.03 for the post takeover and the maximum of 0.07 for pre takeover and 0.06 for the post takeover. The standard deviation of pre is 0.022 and for the post takeover is 0.011. The change of standard deviation shows that company varies in their return on capital employed before the takeover and after the takeover.

Table 5.29: t-test on Leverage ratios

	Total Debt to Equity Ratio	Return on Capital Employed (%)
Mean Correlation	0.23	(0.04)
df (Degree of freedom)	4.00	4.00
SIGN. (2 Tail)	0.08	0.64
t-test	0.09	0.65

Source: t-test Output from SPSS

Table 5.29 shows that there is no significant difference with respect to total debt to equity ratio and return on capital employed percentage of VOLKSWAGEN during pre and post takeover scenario.

Finding on the operational and financial performance of VOLKSWAGEN during its pre and post takeover scenario.

The above results shows that the financial and operational performance of VOLKSWAGEN with respect to its profitability position as well as Leverage position found no significant during pre and post takeover scenario. However, with respect to liquidity position, there is a significant difference between pre and post acquisition scenario.

Based upon the above test results it can be stated that there is an impact of the corporate takeover on the growth and financial performance of the company.

MAHINDRA & MAHINDRA Ltd.

Table 5.30: Operational and Financial performance of MAHINDRA & MAHINDRA Ltd. during the pre and post takeover period

(Rs in crores)

Particulars	2005	2006	2007	2008	2009	2010	2011	2012	2013	2014	2015
	Pre Merger					Year of Merger	Post Merger				
Net Sales	7637	11523	16600	22538	25044	31083	36864	59399	68736	74001	71949
Other Income	1929	1126	1306	2029	1977	720	317	346	389	505	525
Total Revenue	9566	12649	17906	24567	27021	31802	37181	59745	69125	74506	72474
Less: Direct Exp.	(5315)	(7446)	(9891)	(13553)	(14935)	(17394)	(19998)	(35473)	(41844)	(44893)	(42850)
Gross Profit	4251	5203	8015	11014	12086	14408	17183	24272	27281	29613	29624
Less: Indirect and Extraordinary Exp.	(2896)	(3160)	(5136)	(7233)	(8297)	(8415)	(10576)	(16487)	(17324)	(18670)	(20031)
Earnings before Depreciation, Interest & Taxes (EBDIT)	1355	2043	2880	3781	3790	5994	6607	7785	9957	10943	9593
Percentage of EBDIT to Sales	0	0	0	0	0	0	0	0	0	0	0
Less: Interest	(136)	(212)	(283)	(697)	(837)	(1094)	(1135)	(1800)	(2297)	(2954)	(3157)
Less: Depreciation	(239)	(283)	(380)	(582)	(749)	(874)	(972)	(1802)	(2080)	(2170)	(2124)
Profit before Taxes (PBT)	980	1548	2217	2502	2204	4026	4499	4184	5580	5820	4313
Less: Tax	(303)	(403)	(596)	(657)	(542)	(1154)	(1317)	(1408)	(1935)	(1496)	(1720)
Profit after Taxes (PAT)	677	1145	1621	1845	1662	2871	3182	2776	3646	4323	2593
Earnings Per Share (EPS)	59	54	63	66	50	44	53	53	69	79	53

Sources: Data compiled from the official website of Mahindra & Mahindra and Money control

Table no. 5.30 shows pre and post takeover operational and financial performance of MAHINDRA & MAHINDRA during the years from 2005 to 2015. It indicates the fluctuations in PBT (profit before tax), PAT (profit after tax), and EPS (earnings per share) during pre and post takeover scenario.

Table 5.31: Descriptive test statistics based on profitability ratios – pre and post takeover scenario

	Gross Profit Ratio		Net Profit Ratio		Earnings Per Share	
	Pre	Post	Pre	Post	Pre	Post
MAX	0.56	0.47	9.94	8.63	66.00	80.00
MIN	0.45	0.40	6.64	3.60	44.00	55.00
Mean	0.59	0.42	10.52	5.61	63.20	63.20
Std.Dev.	0.04	0.03	1.22	1.88	7.42	11.35

Source: t-test Output from SPSS

Table 5.31 shows that gross profit ratio mean values are 0.59 for pre takeover and 0.42 for the post takeover, with a minimum value of 0.45 for pre takeover and 0.40 for the post takeover, and the maximum value of 0.56 for pre takeover and 0.47 for the post takeover. The standard deviation of pre is 0.04 and for the post takeover is 0.03. The change of standard deviation shows that company varies in their gross profit ratio before the takeover and after the takeover.

Net profit ratio mean values are 10.52 for pre takeover and 5.61 for the post takeover, with a minimum value of 6.64 for pre takeover and 3.60 for the post takeover, and the maximum value of 9.94 for pre takeover and 8.63 for the post takeover. The standard deviation of pre is 1.22 and for the post takeover is 1.88. The change of standard deviation shows that company varies in their net profit ratio before the takeover and after the takeover.

Earnings per share mean values are 63.20 for pre takeover and 63.20 for the post takeover, with a minimum value of 44.00 for pre takeover and 55.00 for the post takeover, and the maximum value of 66.00 for pre takeover and 80.00 for the post takeover. The standard deviation of pre is 7.42 and for the post takeover is 11.35. The change of standard deviation shows that the company varies in their earnings per share before the takeover and after the takeover.

Table 5.32: t-test on Profitability ratios

	Gross Profit Ratio	Net Profit Ratio	Earnings Per Share
Mean Correlation	0.87	0.30	0.87
df (Degree of freedom)	4.00	4.00	4.00
SIGN. (2 Tail)	0.01	0.02	0.18
t-test	0.0001	0.02	0.04

Source: t-test Output from SPSS

Table 5.32 shows that there is a significant difference in gross profit ratio, net profit ratio, and earnings per share during pre and post takeover scenario of MAHINDRA & MAHINDRA. The study concludes that MAHINDRA & MAHINDRA's operational and financial performance has been significantly affected due to its takeover activity.

Table 5.33: Financial position of MAHINDRA & MAHINDRA during pre and post takeover scenario

	2005	2006	2007	2008	2009	2010	2011	2012	2013	2014	2015
	\multicolumn{5}{Pre Merger}	Year of Merger	\multicolumn{5}{Post Merger}
Current Ratio	2.68	2.44	2.22	2.05	1.82	2.11	1.30	1.32	1.30	1.44	1.18
Quick Ratio			1.84	1.64	1.50	1.72	0.97	1.00	0.98	1.13	1.13
Total Debt to Equity Ratio	1.43	1.36	1.08	1.98	1.03	0.90	0.82	0.94	0.96	1.05	0.88
Return on Capital Employed (%)	9.92	12.12	8.81	8.69	6.28	9.16	8.94	7.99	9.13	7.68	9.33

Sources: Data compiled from the official website of Mahindra & Mahindra and Money control

Table no. 5.33 shows MAHINDRA & MAHINDRA's financial position statement for the period 2005 to 2015. It shows that the current ratio of MAHINDRA & MAHINDRA. in the year 2005 is 2.68 which decreased to 1.82 in the year 2009 during pre takeover scenario. During post takeover scenario current ratio of MAHINDRA & MAHINDRA in the year 2011 reported 1.30 which decreased to 1.18

in the year 2015, similarly its return on capital employed percentage in the year 2005 is 9.92 which decreased to 6.28 in the year of 2009 during pre takeover scenario. During post takeover scenario return on capital employed percentage of MAHINDRA & MAHINDRA in the year 2011 reported 8.94 which further increased to 9.33 in the year 2015.

Table 5.34: Descriptive test statistics based on liquidity ratios – pre and post takeover of MAHINDRA & MAHINDRA

	Current ratio		Quick Ratio	
	Pre	Post	Pre	Post
MAX	2.68	1.44	1.84	1.13
MIN	1.82	1.18	1.50	0.97
Mean	2.66	1.31	1.34	1.04
Std. Dev.	0.30	0.09	0.14	0.08

Source: t-test Output from SPSS

Table 5.34 shows that current ratio mean values are 2.66 for pre takeover and 1.31 for the post takeover, with a minimum value of 1.82 for pre takeover and 1.18 for the post takeover and the maximum value of 2.68 for pre takeover and 1.44 for the post takeover. The standard deviation of pre is 0.30 and the post takeover is 0.09. The change of standard deviation shows that the company varies in their current ratio before the takeover and after the takeover.

Quick ratio mean values are 1.34 for pre takeover and 1.04 for the post takeover, with minimum value of 1.50 for pre takeover and 0.97 for the post takeover and maximum value of 1.84 for pre takeover and 1.13 for the post takeover. The standard deviation of pre is 0.14 and for the post takeover is 0.08. The change of standard deviation shows that company varies in their quick ratio before the takeover and after the takeover.

Tables 5.35: t-test on Liquidity ratios

	Current Ratio	Quick Ratio
Mean Correlation	0.22	(0.91)
df (Degree of freedom)	4.00	4.00
SIGN. (2 Tail)	0.00031	0.00018
t-test	0.0030	0.0033

Source: t-test Output from SPSS

Table 5.35 shows that there is a significant difference with respect to the current ratio and quick ratio of MAHINDRA & MAHINDRA during its pre and post takeover scenario. The study concludes that MAHINDRA & MAHINDRA's liquidity position has been significantly affected due to its takeover activity.

Table 5.36: Descriptive test statistics based on leverage ratios – pre and post takeover of MAHINDRA & MAHINDRA

	Total Debt to Equity ratio		Return on capital Employed (%)	
	Pre	Post	Pre	Post
MAX	1.98	1.05	12.12	9.33
MIN	0.90	0.82	6.28	7.68
Mean	1.55	0.93	11.00	8.61
Std. Dev.	0.39	0.08	1.90	0.73

Source: t-test Output from SPSS

Table 5.36 shows that total debt to equity ratio mean values are 1.55 for pre takeover and 0.93 for the post takeover, with a minimum value of 0.90 for pre takeover and 0.82 for the post takeover, and the maximum value of 1.98 for pre takeover and 1.05 for the post takeover. The standard deviation of pre is 0.39 and for the post takeover is 0.08. The change of standard deviation shows that the company varies in their total debt to equity ratio before the takeover and after the takeover.

Return on capital employed percent mean values are 11.00 for pre takeover and 8.61 for the post takeover, with a minimum value of 6.28 for pre takeover and 7.68 for the post takeover and the maximum value of 12.12 for pre takeover and 9.33 for the post takeover. The standard deviation of pre is 1.90 and for the post takeover is 0.73. The change of standard deviation shows that the company varies in their return on capital employed percentage before the takeover and after the takeover.

Table 5.37: t-test on Leverage ratios

	Total Debt to Equity Ratio	Return on Capital Employed (%)
Mean Correlation	0.57	(0.55)
df (Degree of freedom)	4.00	4.00
SIGN. (2 Tail)	0.03	0.60
t-test	0.04	0.66

Source: t-test Output from SPSS

Table 5.37 shows that there is a significant difference with respect to total debt to equity ratio of MAHINDRA & MAHINDRA. However, in case of return on capital employed percentage of MAHINDRA & MAHINDRA, there is no significant difference during pre and post takeover scenario.

Finding on the operational and financial performance of MAHINDRA & MAHINDRA during its pre and post acquisition scenario.

The above results show that the financial and operational performance of MAHINDRA & MAHINDRA with respect to its profitability and liquidity position there is a significant difference between pre and post acquisition scenario. Further, the result of leverage ratios of MAHINDRA & MAHINDRA also found significant during pre and post takeover scenario except its return on capital employed percentage.

Based upon the above test results it can be concluded that there is an impact of the corporate takeover on the growth and financial performance of the company.

DAIMLER CHRYSLER

Table 5.38: Operational and Financial performance of DAIMLER CHRYSLER during the pre and post merger period

(in Millions €)

Particulars	1995	1996	1997	1998	1999	2000	2001	2002	2003
		Pre Merger		Year of Merger			Post Merger		
Sales Revenues	52,522	101,415	117,572	131,782	149,985	162,384	152,873	147,368	136,437
Financial Income, net	474	408	633	763	333	156	154	2,206	(2,816)
Other income	888	848	957	1,215	827	946	1,212	700	1,579
Net Revenues	53,885	102,671	119,162	133,760	151,145	163,486	154,239	150,274	135,200
Cost of sales	(44,210)	(78,995)	(92,953)	(103,721)	(119,688)	(134,370)	(128,394)	(119,624)	(109,926)
Gross margin	9,675	23,676	26,209	30,039	31,457	29,116	25,845	30,650	25,274
% of Gross margin to Net Revenue	21%	23%	22%	26%	20%	16%	15%	21%	19%
Indirect Expenses	(13,364)	(17,983)	(20,029)	(21,885)	(21,800)	(24,640)	(27,328)	(24,817)	(23,847)
Extraordinary & other indirect expenses	(59)	(124)	(115)	(259)	622	5,417	44		
Income before income taxes (EBT)	(3,748)	5,569	6,065	7,895	10,279	9,893	(1,439)	5,833	1,427
% of EBT to Net Revenue	(7%)	5%	5%	6%	7%	6%	(1%)	4%	1%
Less: Taxes	826	(1,547)	482	(3,075)	(4,533)	(1,999)	777	(1,115)	(979)
Net income	(2,922)	4,022	6,547	4,820	5,746	7,894	(662)	4,718	448
% of Net Income to Net Revenue	8%	4%	5%	7%	2%	3%	1%	2%	0%
Basic earnings per share	(5.70)	2.09	3.90	4.03	7.78	8.87	5.66	5.68	6.44

Sources: Data compiled from the official website of Daimler AG and Money control

Table no. 5.38 shows pre and post takeover operational and financial performance of DAIMLER CHRYSLER during the years from 1995 to 2003. It indicates the fluctuations in PBT (profit before tax), PAT (profit after tax), and EPS (earnings per share) during pre and post takeover scenario.

Table 5.39: Descriptive test statistics based on profitability ratios – pre and post takeover scenario

	Gross Profit Ratio		Net Profit Ratio		Earnings Per Share	
	Pre	Post	Pre	Post	Pre	Post
MAX	0.26	0.21	0.08	0.03	4.03	8.87
MIN	0.21	0.15	0.04	0.01	(5.70)	5.66
Mean	0.23	0.18	0.06	0.02	1.08	7.00
Std. Dev.	0.02	0.03	0.02	0.01	4.60	1.60

Source: t-test Output from SPSS

Table 5.39 shows that gross profit ratio mean values are 0.23 for pre takeover and 0.18 for the post takeover, with a minimum value of 0.21 for pre takeover and 0.15 for post takeover, and the maximum value of 0.26 for pre takeover and 0.21 for post takeover. The standard deviation of pre is 0.02 and for post takeover is 0.03. The change of standard deviation shows that company varies in their gross profit ratio before takeover and after takeover.

Net profit ratio mean values are 0.06 for pre takeover and 0.02 for the post takeover, with a minimum value of 0.04 for pre takeover and 0.01 for the post takeover, and the maximum value of 0.08 for pre takeover and 0.03 for post takeover. The standard deviation of pre is 0.02 and for the post takeover is 0.01. The change of standard deviation shows that company varies in their net profit ratio before takeover and after takeover.

Earnings per share mean values are 1.08 for pre takeover and 7.00 for the post takeover, with a minimum value of (5.70) for pre takeover and 5.66 for post takeover and the maximum value of 4.03 for pre takeover and 8.87 for post takeover. The standard deviation of pre is 4.60 and for the post takeover is 1.60. The change of standard deviation shows that company varies in their earnings per share before takeover and after takeover.

Table 5.40: t-test on Profitability ratios

	Gross Profit Ratio	Net Profit Ratio	Earnings Per Share
Mean Correlation	0.41	(0.22)	(0.50)
df (Degree of freedom)	3.00	3.00	3.00
SIGN. (2 Tail)	0.124	0.12	0.1241
t-test	0.04	0.03	0.049

Source: t-test Output from SPSS

Table 5.40 shows that there is a significant difference in gross profit ratio, net profit ratio, and earnings per share during pre and post takeover scenario of DAIMLER CHRYSLER. The study concludes that DAIMLER CHRYSLER's operational and financial performance has been significantly affected due to its takeover activity.

Table 5.41: Financial position of DAIMLER CHRYSLER during pre and post takeover scenario

	1993	1994	1995	1996	1997	1998	1999	2000	2001	2002	2003
	Pre Merger					Year of Merger	Post Merger				
Current Ratio	0.85	0.86	1.00	1.28	1.34	1.44	1.57	1.76	1.70	1.62	1.76
Quick Ratio	0.55	0.57	0.68	1.02	1.03	1.12	1.23	1.36	1.31	1.23	1.34
Total Debt to Equity Ratio	0.39	0.30	1.27	1.06	1.20	1.30	1.76	1.98	2.31	2.24	2.17
Return on Capital Employed (%)	50.09	61.44	59.27	80.73	39.35	39.91	29.76	21.98	18.92	25.46	21.79

Sources: Data compiled from the official website of Daimler AG and Money control

Table no. 5.41 shows the DAIMLER CHRYSLER's financial position statement for the period 1993 to 2003. It shows that current ratio of DAIMLER CHRYSLER in the year 1993 is 0.85 which increase to 1.34 in the year 1997 during pre takeover scenario. During post takeover scenario current ratio of DAIMLER CHRYSLER in the year 1999 reported at 1.57 which increased to 1.76 in the year 2003. Similarly its return on capital employed in the year 1993 is 50.09 which decreased to 39.35 in the year of 1997 during pre takeover scenario. During post takeover scenario return on capital employed of DAIMLER CHRYSLER in the year 1999 reported at 29.76 which further decreased to 21.79 in the year 2003.

Table 5.42: Descriptive test statistics based on liquidity ratios – pre and post takeover of DAIMLER CHRYSLER

	Current ratio		Quick Ratio	
	Pre	post	Pre	post
MAX	1.34	1.76	1.03	1.36
MIN	0.85	1.57	0.55	1.23
Mean	1.07	1.68	0.77	1.29
Std.Dev.	0.23	0.08	0.24	0.06

Source: t-test Output from SPSS

Table 5.42 shows that current ratio mean values are 1.07 for pre takeover and 1.68 for the post takeover, with a minimum value of 0.85 for pre takeover and 1.57 for the post takeover and the maximum value of 1.34 for pre takeover and 1.76 for the post takeover. The standard deviation of pre is 0.23 and post is 0.08. The change of standard deviation shows that the company varies in their current ratio before the takeover and after the takeover.

Quick ratio mean values are 0.77 for pre takeover and 1.29 for post takeover, with a minimum value of 0.55 for pre takeover and 1.23 for post takeover, and the maximum value of 1.03 for pre takeover and 1.36 for the post takeover. The standard deviation of pre is 0.24 and for the post takeover is 0.06. The change of standard deviation shows that the company varies in their quick ratio before the takeover and after the takeover.

Tables 5.43: t-test on Liquidity ratios

	Current Ratio	Quick Ratio
Mean Correlation	0.21	(0.10)
Df (Degree of freedom)	4.00	4.00
SIGN. (2 Tail)	0.0005	0.0014
t-test	0.0037	0.0094

Source: t-test Output from SPSS

Table 5.43 shows that there is a significant difference with respect to current ratio and quick ratio of DAIMLER CHRYSLER during its pre and post takeover scenario. The study concludes that DAIMLER CHRYSLERs liquidity position has been significantly affected due to its takeover activity.

Table 5.44: Descriptive test statistics based on Leverage ratios – pre and post takeover of DAIMLER CHRYSLER

	Total Debt to Equity ratio		Return on capital Employed (%)	
	Pre	Post	Pre	Post
MAX	1.27	2.31	80.73	29.76
MIN	0.30	1.76	39.35	18.92
Mean	0.84	2.09	0.58	0.24
Std.Dev.	0.46	0.22	0.15	0.04

Source: t-test Output from SPSS

Table 5.44 shows that total debt to equity ratio mean values are 0.84 for pre takeover and 2.09 for the post takeover, with a minimum value of 0.30 for pre takeover 1.76 for the post takeover and the maximum value of 1.27 for pre takeover and 2.31 for the post takeover. The standard deviation of pre is 0.46 and post is 0.22. The change of standard deviation shows that the company varies in their total debt to equity ratio before the takeover and after the takeover.

Return on capital employed percentage mean values are 0.58 for pre takeover and 0.24 for the post takeover, with a minimum value of 39.35 for pre takeover and 18.92 for the post takeover and the maximum value of 80.73 for pre takeover and 29.76 for the post takeover. The standard deviation of pre is 0.15 and for the post takeover is 0.04. The change of standard deviation shows that the company varies in their return on capital employed percentage before the takeover and after the takeover.

Tables 5.45: t-test on Leverage ratios

	Total Debt to Equity Ratio	Return on Capital Employed (%)
Mean Correlation	0.88	0.06
df (Degree of freedom)	4.00	4.00
SIGN. (2 Tail)	0.001	0.0012
t-test	0.006	0.008

Source: t-test Output from SPSS

Table 5.45 shows that there is a significant difference with respect to total debt to equity ratio and return on capital employed percentage of DAIMLER CHRYSLER during the pre and post takeover scenario.

Finding on the operational and financial performance of DAIMLER CHRYSLER during its pre and post takeover scenario.

The above results show that the financial and operational performance of DAIMLER CHRYSLER with respect to its profitability, liquidity, and leverage position, there is a significant difference between pre and post takeover scenario.

Based upon the above test results it can be concluded that there is an impact of the corporate takeover on the growth and financial performance of the company.

5.3 Hypothesis testing and Results

Based on the review of literature and experience survey, the following hypothesis tested under the study.

The following table 5.46 shows the t-test results of the pre and post-acquisition performance of four selected companies:

Table 5.46: t-test summary of Profitability indicators

S. No.	Name of Company	Gross Profit Ratio	Net Profit Ratio	EPS
1.	TATA Motors Ltd.	0.04	0.03	0.04
2.	Volkswagen	0.75	0.20	0.89
3.	Mahindra & Mahindra Ltd.	0.001	0.02	0.04
4.	Daimler Benz	0.04	0.03	0.049

The above table 5.46 shows that after applying t-test there is a significant relationship in profitability indicators of three companies, out of selected four companies, as $P > 0.05$ (less than 0.05).

Therefore, on the basis of the above analysis, it can be concluded that there is an impact on the profitability of the selected companies, between the pre and post acquisition scenario of selected companies.

The following liquidity indicators are considered for study:

i. Current Ratio

ii. Quick Ratio

Table 5.47 shows the t-test results of the pre and post acquisition performance of four selected companies with respect to liquidity position:

Table 5.47: t-test summary of Liquidity indicators

S. No.	Name of Company	Current Ratio	Quick Ratio
1.	TATA Motors Ltd.	0.024	0.009
2.	Volkswagen	0.01	0.03
3.	Mahindra & Mahindra Ltd.	0.003	0.0033
4.	Daimler Benz	0.0037	0.0094

The above table 5.47 shows that after applying t-test there is a significant relationship in liquidity indicators of all the selected four companies, as $P > 0.05$ (less than 0.05).

Therefore, on the basis of the above analysis, it can be concluded that there is an impact on the liquidity position of the selected companies, between the pre and post acquisition.

The following leverage indicators are considered for study:

i. Total Debt to Equity Ratio

ii. Return on Capital Employed percentage (ROCE)

The table 5.48 shows the t-test results of the pre and post acquisition performance of four selected companies:

Table 5.48: t-test summary of Leverage indicators

S. No.	Name of Company	Total Debt to Equity Ratio	Return on Capital Employed percentage
1.	TATA Motors Ltd.	0.29	0.32
2.	Volkswagen	0.09	0.65
3.	Mahindra & Mahindra Ltd.	0.04	0.66
4.	Daimler Benz	0.006	0.008

The above table 5.48 shows that after applying t-test there is a significant relationship in debt to equity ratio between the three companies, out of selected four companies as $P > 0.05$ (less than 0.05). Also in case of return on capital employed there is a significant relationship in one company, out of selected four companies as $P > 0.05$ (less than 0.05).

Therefore, on the basis of the above analysis, it can be concluded that there is an impact on the leverage position of the selected companies, between the pre and post acquisition.

5.4 Results of Factor Analysis

On the basis of factor analysis, the following results have been derived:

i.) For a successful takeover necessary factors are-a new segment of customers, the greater fuel efficiency of vehicles and better return on equity i.e wealth maximization.

ii.) Growth drivers of the takeover are-operating synergy increased buying power of the people and popularity of two-wheelers and passenger cars, subject to the management of corporate cultural differences.

iii.) The growth drivers of the automobile industry include-the availability of skilled labor at low cost, Robust Research and Development centers, and greater fuel efficiency.

iv.) Proper communication among the stakeholders during the corporate transition period, the intent of geographical expansion and sharing of information about the future growth potentials are important reasons behind successful corporate takeovers.

v.) The primary reasons for corporate takeover are to improve the performance of the company and to improve the interest of its stakeholders.

Further, following the major key factors have been identified to motivate selected companies to go for the takeover of target companies:

1. **Value of Company:** study reveals that the strongest motivating factor behind an M&A activity is a financial benefit. The value maximizing states that an acquisition

aims at maximizing the combined market worth of the partnering organizations. Hence, there is a rise in the market worth of both the acquirer and the target companies. This then is 'synergy,' wherein the higher value is created and better resource management is achieved. There are five main stages of market activity - market power, economies of scale and economies of scope, co-insurance, and financial diversification. Horizontal acquisitions aim at decreasing competition, thereby increasing profits and enhancing the worth of the bidder. The synergy created by the economy of scale is possible in the case of horizontal mergers or if the merging entities share similar production factors. Economies of scope can be achieved in case a company acquires a related company (not operating in the same market). Such synergy is best exemplified in the case of vertical acquisition wherein the raw material of one company is actually the final output of another company. Now, if two such companies are amalgamated, not only the sales will increase but the production costs will go down, too, thereby maximizing profits. Another post acquisition benefit is the enhanced level of debt capacity. This increases the market worth of the acquiring company even more. The commonest motive for takeovers is, indeed, value maximisation. The fundamental idea is that the combined value of the acquiring and the target company is more in value than the sum of the individual values of the two companies. With the amalgamation of production and infrastructure costs and gaining higher efficiency and market power, a takeover bid is ultimately expected to aid economies of scale, thereby increasing the value of the company.

2. **Managing growth during the transition**: The study reveals that after takeover it is necessary to have a well-imagined implementation plan in place to ensure that the integration of the strengths of the merging parties is smooth. This must include the anticipation of eventualities and due consideration for personal and soft elements. Mergers involving different geographies may come with socio-cultural challenges wherein it becomes difficult for the different workforces to perform as a team because of personal frictions. Even if geographies and social norms are same, sometimes the corporate culture and working style may differ so much that the employees of the newly merged entity find it impossible to get together as a team. Either case, such dis-integration severely compromises on efficiency, reducing productivity and leading to

loss and final failure of the merger. Such issues need to be addressed adequately during the transition period.

The decision to takeover another business company is a very crucial one. and must be taken after a detailed and exhaustive understanding of all the underlying factors. Given that acquisitions have a long history, there must be tangible merits to it.

3. **Growth Driver**: The study reveals that the major catalyst that spurs potential M&A deal is the attraction of inorganic growth. The other option of growing by itself is lengthy and ridden with more regulatory and compliance pressures. The other catalysts are likely to be the availability of a worthy target company available for relatively low prices (distress sale), or the acquiring company's will to reduce dependence on vendors by acquiring them through forward or backward integration. Some of the benefits of triggers for any M&A are listed below:

i. Proportionate saving in cost.
ii. Synergistic operations and increased efficiency.
iii. Access to a different market, be it products, be it new geographies or new stream.
iv. Access to foreign fund.
v. Upgraded technology.
vi. Garnering market share.

4. **Future plans and geographical expansion**: The study also reveals that M&A deals are undertaken with every hope of business expansion, better profitability, and reduced costs. These are massive decisions and require detailed studies with respect to potential impacts on markets and stakeholders. It makes sound business sense to have diversity in the products and services by the acquisition of the products and services of a company available for sale. If profits dip in one category of product or service, it can rise in another so that overall the company remains profitable. With a takeover, it becomes possible for one company to acquire a different type of product or service being offered by the target company.

5. **Shareholders Interest:** The study further reveals that a takeover impacts both acquiring and target Company's share price. The shareholder's wealth in value change or abnormal return is expected. Before acquiring another company, investors in a company have to decide whether the acquisition will be beneficial to them or not. Therefore, they must determine the worth of the target company.

M&A provide access to pooled resources, better infrastructure, assets base, skill, expertise, market-based, and so on; these pooled resources yield better performance. Declined profitability in relation to pooled resources, contrary to the expected improvement, reflects underutilization of the inherent potential of M&A by the acquirer companies. This necessitates an effective integration approach; proper assessment of pooled assets, resources and opportunities offered by M&A, in conjunction with pre-planned integration strategy.

CHAPTER 6
CONCLUSION, RECOMMENDATIONS, AND SCOPE FOR FUTURE RESEARCH

This chapter gives an overview of the whole of the study undertaken to examine the pre and post merger impact of takeovers on selected companies under the automobile industry. The study has been conducted based on major objectives with respect to review the past and present scenario of merger and takeover practices in US, UK, and in India, to examine the impact of the corporate takeover on the growth and financial performance of the selected companies under the automotive industry, to investigate the growth drivers and challenges for the automotive industry, and to identify the future trend of the automotive industry, in India.

6.1 Conclusion

While some studies have already been done on the impact of M&A, this study aims to analyze the impact of the corporate takeovers specifically on the automobile industry. This study by recapitulating the theoretical underpinnings of the tests conducted, recent empirical research in the automotive takeover and acquisition suggests that each M&A deal is individualistic, having its own set of issues and problems. However, the underlying principle does not vary that two business organizations decide to merge, getting consolidated into one entity, with the aim of mutual financial or strategic benefits. It is, however, a truly complicated process. Testing for the existence of this relationship provides evidence on the extent to which merger and acquisition affect a company.

The study reveals that there are cases of failure of takeovers in the automotive industry also. The most critical factor to ensure the success of a deal is a well-executed integration plan. Other key elements include accurate and ethical valuation, an effective due diligence plan and the general economic environment. The company's ability to recognize cultural differences is also an important factor. Organizations opting for a merger are likely to have distinct working cultures, and as such, any merger plan must aim at a careful, seamless integration. In some cases, when organizations undertaking mergers and acquisitions are geographically distant, then it may be beneficial to keep these two entities functionally distinct.

M&A can be a strategically sound move to stabilize one's base and grow revenues, though it does require a stable economy. M&A deals require detailed strategic thinking and execution. Senior management involvement at all levels is strongly required.

Based on the objectives set it can be further concluded as under:

Past and present scenario of M&A deals in the automotive industry: Starting in the 1890s, the USA M&A deals can be divided into six distinct phases or waves. While horizontal mergers were the commonest ones in the first wave (1895-1904), the second wave was qualified by vertical M&As (1916-1929). Conglomerate mergers characterized in the third wave that lasted 1965-1969. The fourth wave (1981-1989) witnessed a number of hostile takeovers. Riding on the wings of growing globalization, the fifth wave showcased a number of cross-border mergers (1990-2000). The wave currently going on (2003-present) has shown evidence of shareholder activism and leveraged buy-outs.

Barring two mini merger booms that happened in the 1890s and 1920s, the first merger wave in the UK may be traced to 1968. Here too, the horizontal mergers took precedence, largely sponsored by Britain's Industrial Reorganization Corporation. The second wave was also one characterized largely by the horizontal mergers, through a few conglomerate mergers did occur. The third wave of the 1980s was the most substantial one, riding on the wave of the stock market bull run as well as the 'Big Bang Deregulation' of the financial services sector in London. The fourth wave was recorded in the 1990s, with privatizations and deregulations being the striking features driving the deals.

Seeds of the Indian automotive industry were sown in the 1880s, but the wave of M&A's came nearly 100 years later, in the 1980s. The acquisitions were usually friendly ones, negotiated through mutual dealings. The Indian automotive industry showed significant advances since de-licensing and opening up of the sector to the abolition of licensing in 1991.

Takeover practices worldwide: The study reveals that horizontal, vertical and conglomerated acquisitions are three types of takeover transactions by way of which a bidder company acquires control over a target company. A horizontal acquisition refers to two companies, performing and competing in the same industry, though, perhaps in geographically separate regions. It leads to competition elimination, new market share, and economies of scale. In the case of a vertical acquisition, the major aim behind acquiring company's desire to have complete ownership over the entire production chain, thereby making its market position secure. The conglomerated acquisitions are aimed at risk-dilution, and often those companies opt for it, belongs to the high-risk industry. Their prime aim is not to have all one's eggs in the same basket and try to obtain profits from a wide range of industries.

Impact of corporate takeovers on the growth and financial performance of selected companies under the automotive industry: In the present study four companies selected from the automobile industry which has kept their identities, pre and post-merger. After applying statistical tests it is found that there is a significant relationship in profitability indicators, liquidity indicators and leverage indicators of the selected companies. Therefore, it is concluded that there is a significant (pre and post-acquisition) impact of the corporate takeover on the growth and financial performance of the selected companies under the automotive industry.

Based on the study and test conducted it can be concluded that after the acquisitions, since the difference in variations is marginal, their profitability, liquidity and leverage positions have been improved.

Based on the study of Daimler and Chrysler merger it can be stated that this was not much effective mainly due to cultural differences among two organizations. Therefore, it needs to be appropriately addressed in all future mergers. Apart from this one must consider upgradation of technology with know-how collaboration, improving shareholders involvement, geographical and market expansion.

Growth Drivers and Challenges: To achieve this objective, the growth drivers are derived through primary data collected with the help of questionnaires and tested through factor analysis.

On the basis of factor analysis it can be stated that the following five components are major factors affecting mergers and acquisitions of companies under automobile industry:

i. For a successful takeover, necessary factors are - a new segment of customers, greater fuel efficiency of vehicles and better return on equity.

ii. Growth drivers of the takeover are-operating synergy increased buying power of people and popularity of two-wheeler and passenger cars.

iii. The growth drivers of the automobile industry include-the availability of skilled labor at low cost and robust research and development centers.

iv. Proper communication among the shareholders during the corporate transition period, the intent of geographical expansion and sharing of information about the future growth potentials are important reasons behind successful corporate takeovers.

v. The primary reasons for corporate takeover are to enhance the performance of the company and to increase the interest of its stakeholders.

Future Trend: After studying the past and present scenario of the automobile industry, performance indicators of selected companies and growth drivers of the automobile industry, the following future trend is observed:

i. Mergers and acquisitions are significant components of inorganic growth, and despite their undeniable complexities and some notable failures, they are likely to remain so in the future. This is especially so for the automotive sector.

ii. In India, such activities are currently on the ascent, given its vibrant market, young labor force and low manufacturing costs. Availability of liquidity, desire for strategic initiatives, globalization and technological leaps in the automotive industry are the important factors taking M&A deals onwards. Stress must be on adherence to applicable legislation, due diligence of target company and above all, on usually ignored socio-cultural factors and the human element.

iii. Given its years' old history, merger and acquisition deals are assuredly time-tested. Economies of scale, synergies, enhancement of productivity and profitability, reduction of risk and elimination of competition are well-proven positive after-effects of the merger deals. Another spur for takeover is also the desire to possess the state of the art technologies available, such as online cab services, automation, and smart fuel engines which are helping to meet the novel challenges posed by today's environment. With globalization and IT developments, cross-border deals and conglomerate deals are becoming the flavor of the season.

iv. The automotive industry in India is one of the most progressive passenger cars market across the globe, the largest motorcycle manufacturer, the second-largest two-wheeler manufacturer, and the world's fifth-largest commercial vehicle manufacturer, apart from becoming a pivotal export center for special utility vehicles. The top global automotive manufacturers hope to reap the benefits of India's economic and inexpensive manufacturing practices. In fact, it is possible that in near future India becomes the hub of vehicle manufacture and supply, fulfilling the world's need of the vehicles.

v. India is perfectly poised to take the next big leap in the automotive market. Not only as a rich market, not only with its vibrant young workforce, not only as a target company but as an international acquirer. The takeover of Jaguar Land Rover by Tata Motors is a prime example of the same. Business-friendly policies such as "Make in India" and improved "Ease of doing business" are bound to show their positive impacts in the coming years.

6.2 Recommendations

It is important to conduct in-depth analysis and due-diligence before proceeding for an M&A. Various departments of a company, like, finance, human resource and legal departments, collect information in order to analyze factors that can help in assessing and determining the success of an M&A. These include factors such as rise or decline of revenues, profits, productivity, market share, share prices and so on.

Following aspects observed significant to be kept in mind while undertaking any M&A activity:

i. Identifying opportunities for providing competitive benchmarking measurements for fast improvement in the company.
ii. Measuring market and technology trends to forecast the future growth potential of a company's technology.
iii. Personal ambition/motives of industry and financial reward, which the company will earn due to M&A, should be targeted.
iv. Big takeovers attract media and boost brand image, but companies should link reward to growth. Peer pressure and pressure from advisers and media to opt for takeovers should be taken care of.
v. Concern that the company may be left behind or over-confidence should be taken care of.
vi. Always make use of surplus cash, high share price, bargain hunting, and asset stripping.
vii. Motive should be to extend the business locations, markets, and globalization.
viii. Improving negotiation and price by ascertaining the company's exact market position.
ix. Providing credibility and assurance to investors and bankers.
x. Gauging the customer's attitudes towards the brand image.
xi. Business houses in India, with their remarkable trend of maintaining stability when the world economy trembles, have been front-ending global and domestic M&A players.

As the industry recovers, automobile companies across the value chain must focus on:

i. profitability and sustainable growth,

ii. financial and operational flexibility,

iii. investments in new technologies, and

iv. seizing opportunities in high-growth markets.

Further, the study suggests that following globally tried-and-tested best practices aimed at easing out the trickiest challenges in the field of M&A. Of course, not all pointers shall be relevant for each case, hence it is recommended to select the approaches most pertinent to respective situations.

i. **Planning of integration should start at the earliest:** Waiting for the eleventh hour to start integration planning will be a real mistake and to be avoided at all costs. In fact, this planning can begin even before due diligence has started.

ii. **Expecting no glitches is unrealistic**: There will indeed be some glitch or the other, in this complicated process. The important point is to accept the same and discuss its smoothing over.

iii. **Hiring a professional consulting agency:** Such professional assistance can go a long way in smoothening an otherwise complicated process.

iv. **Assessment of competitors and non-competitors:** External benchmarking of competitors as well as non-competitors is of prime importance especially with the aim of incorporating industry's best practices within the newly combined organization's code.

v. **Formation of integration execution task group, involving important personnel from the participating organizations:** The key personnel involved in the decision-making process also to be a part of the said task group, interacting with the others and collecting constructive feedback.

vi. **Due importance to the individual organization's culture:** Cultural features may appear insignificant in relation to the financial and managerial ones but are critical nevertheless, and the wise shall not ignore them.

vii. **Identifying the best practices of individual organizations:** It is observed that the larger acquiring company's business practices will supersede the target company. However, it is quite possible that the target company, the one being acquired, follows certain business practices that are indeed good enough to be adopted by the newly merged entity. Attention and due scrutiny are, therefore, recommended for this.

viii. **Involvement of the HR department and middle-tier employees:** Timely involvement of the HR right from the beginning of the decision-making process is crucial to the success of any M&A process. They are more approachable individuals to whom the regular employees will turn to, for answers and doubt-clarification.

ix. **Keeping the employees up-to-date with all aspects of the M&A:** The mode of communication can be more than one and its frequency can be high, for the information to be truly assimilated by the employees. Intranet, newsletter publications, home-delivered leaflets, and group interactions are a few commonly availed methods of putting the information across. The who, what and how of communication is of extreme importance.

x. **Careful selection of new staff:** The merged company is a new entity and it is not necessary that employees of the target or acquirer companies will be the most suitable. Adequate attention must, therefore, be paid to employee selection.

xi. **Understanding when to 'Just say No':** Some mergers may just be plumb impossible, it is necessary to understand such situations where the cultural disparities are too intense to allow for a seamless merger.

6.3 Limitations

The present study has been undertaken subject to the following limitations:

i. The study is based on primary as well as secondary data and therefore the quality of the study depends purely upon the accuracy, reliability, and quality of data provided by the sources.

ii. The study relates to the impact of the corporate takeover on the automobile industry only.

iii. The scope of this study is wider but the sample size is limited.

iv. There are different methods to measure efficiency, effectiveness, and profitability. Moreover, some tests cannot be expressed in monetary terms.

v. The individual effort will be limited so it is also a limitation of the study.

vi. The study is limited to a specific period of time.

6.4 Scope for Future Research

The present study reveals that the rise in the number of takeover and acquisitions in the automotive industry could be a result of the increasing severe consolidation pressures. Takeover and acquisitions take several forms and occur for different reasons. In the past three decades the automotive industry has witnessed intense pressure and challenges. Due to extreme competition, firms are operating at minimum efficiency, making it difficult for the plant operations to further reduce costs. Cost savings have been observed to be primarily from the areas like R&D. Such costs play a significant role today and in near the future mergers and takeovers as the emphasis on innovations is rising. Another distinct automotive industry specific feature is that the firms are inclined towards making takeover decisions of acquiring products or making entry into new markets effectively and speedily, rather than operating on their own. Therefore, it has acted as a good proxy, as many companies in various industries re seeking growth and strategic gains through acquisitions. Therefore, future research should be done in areas to identify the reasons of failures in acquisitions and to zero down the measures to eliminate such reasons.

The scope of future research may lie in the areas where the acquisition has failed and reasons thereof and identify the measures to mitigate those reasons.

The future researchers may also undertake the study on the same lines for analyzing the impact of corporate takeovers on other industries such as electrical appliances, heavy engineering goods, service sector, financial sector and many more.